GARGOYLE COUNTRY

The Inspiring Geology of Springfield & Greene County

by Jerry D. Vineyard

Photography by the author unless otherwise credited.

Funding support provided by the Southeast Rotary Club of Springfield, Greene County, City of Springfield Department of Public Works Stormwater Division, Springfield Underground, and Watershed Committee of the Ozarks.

Northview Formation in road cut, Hwy. 65 North.

Contents

Stockpiles of crushed stone at Springfield Underground, Hwy. 65 and Division Street.

Weathered hay barn at Lost Hill Park sits on alluvial deposits. Once common, hay barns are rapidly disappearing from the Greene County landscape.

Concrete slab low-water bridge across the South Dry Sac, Lost Hill Park. Increasingly rare, such crossings are often hazardous when drivers misjudge the depth and power of the current.

Landscaping stones add character and texture to the urban scene.

Foreword

Dr. Kenneth C. Thomson, a distinguished professor of geology and surveying at Southwest Missouri State University (now Missouri State) came to Springfield from his home state of Utah. He brought with him a disciplined intellect and an engaging personality. While he always yearned to return to Utah, he and his wife Edda Mae made Springfield their home and raised their family here.

Southwest Missouri was greatly different from Utah, so Dr. Thomson set about studying the unfamiliar geology in a professional manner, believing the best way to understand something was to experience it. Embracing the region's karst topography, Dr. Thomson and his students mapped caves, investigated springs, and documented sinkholes. He personally mapped the geology of Greene County, and became the area's best-known and most trusted authority on the region's geology. In a very real sense, he was the successor to another Springfield geologist, Dr. Edward M. Shepard, who in the nineteenth century was an internationally-known professor at Drury College and a champion of the geology of Springfield and Greene County.

Author Vineyard with his field assistant, getting the word straight from the horse's mouth. Photo by Jack Ray.

Dr. Thomson was at the height of his career when both he and his wife were killed in a tragic car accident in 2004. They were en route home from a trip to Utah to attend the wedding of their daughter when the accident happened. After his death, Dr. Thomson's children offered his papers, professional library and his geological collections to the Watershed Committee of the Ozarks, to be used for the benefit of Springfield-area citizens, in honor of the mutual respect Dr. Thomson and the people of this area had developed over his long career. These materials now reside in the Missouri Institute of Natural Science, where they are available for

Dr. Kenneth C. Thomson with Jerry D. Vineyard.

public use. The collection includes literally hundreds of cave maps, reports, and photographs, consulting reports, and an extensive library related to karst geology. The availability of this collection, amassed during a lifetime of teaching and professional activity, constitutes Dr. Thomson's posthumous legacy to the people of southwest Missouri.

The Watershed Committee published Dr. Thomson's "Geology of Greene County" in 1986. The targeted audience was geologists and other professionals who would be conversant with geologic concepts and terminology. There were no photographs and no color illustrations, except for the geologic map. The current edition, by Jerry D. Vineyard, targets a broader readership, using numerous illustrations to highlight the region's interesting geology. The Geologic Map of Greene County, however, is reproduced at a smaller scale, without significant changes, because post-1986 geologic mapping has changed few details since Thomson's ground-breaking work more than 20 years ago.

Finally, there are added features on cryptoexplosive structures, water resources, a greatly expanded section on karst, and an extensive section on where to see and experience geology in the field.

Introduction

The statement, "Springfield and Greene County are geologically-friendly places" becomes obvious when one considers all of the geology-related characteristics of what we call our "home country."

Springfield-Greene County is a geology-friendly place. All the major roads leading into the county from the north, south, east and west cut through bedrock, leaving exposures of the area's defining geology for all to enjoy. I-44 and Hwys. 13, 60 and 65, for example, are high-speed, four-lane routes that offer smooth passage and scenic views to travelers. The unusual gargoyle-like exposures of the Burlington Limestone that flank parts of these roads hint at the special geology that makes Springfield-Greene County such a wonderful place to live and work. The name "Gargoyle Country" focuses attention on the erosional surface of the Burlington Limestone, named for exposures in Burlington, Iowa. Had the geologist who named the formation been in Springfield at the time, it is a safe bet that we would now be calling it the "Springfield Limestone." Because some 65 percent of the land surface in Greene County is underlain by this remarkable limestone that erodes into gargoyle-like forms, it is no accident that area residents like to place fancifully-shaped limestone rocks around their homes and workplaces. In Springfield, it is literally true that you can have "a piece of the rock."

Roadcuts are said to be the geologist's best friend—especially in

the Midwest, where fields and forests tend to hide the bedrock. In Springfield-Greene County, highway engineers wisely chose to leave bedrock exposed where possible, to add character and scenic diversity to the landscape. Even the name—Springfield—suggests a fascinating place where springs abound, and so it is. The city was established partly because it offered springs for water supply. Today, Fulbright Spring remains an integral part of Springfield's multi-source public water system.

"Jabba the Hut" gargoyle, Hwy. 160 West.

Limestone is the primary rock type in Springfield-Greene County, either as calcium carbonate (limestone), or calcium magnesium carbonate (dolomite). Carbonate rocks are especially vulnerable to solution by groundwater, so areas having limestone bedrock tend to develop a distinctive landscape called "karst." Karst develops gradually over time, producing a landscape rich in sinkholes, springs, caves, and losing streams, precisely the landscape that makes this area so attractive. Karst does, however, present challenges to construction projects of all kinds, but it repays care in construction with opportunities to make geologic features an added bonus to a region already rich in diversity.

Landscaping stones of Burlington Limestone grace many home sites.

Limestone is an especially useful rock, the essential material for making cement, the aggregate for making concrete, and the raw material for quicklime, mortar, and plaster. Springfield's fortuitous location atop strata of high-quality limestone helps to keep construction costs in check, while the underground space resulting from initial limestone mining is used for a wide variety of purposes including warehousing, both hot and cold-storage, records management, and manufacturing. Even the weathered, uneven surface of limestone formations is useful, for it produces attractive landscaping stones that are used for architectural accents throughout the region.

Past mineral production in Greene County has included lead and zinc, iron, and coal, though none of these commodities were ever produced in great quantities. Nevertheless, they add variety and interest to the area's geologic history. While it is

Inside Riverbluff Cave. The cave holds a treasure trove of fossils, found in sediment exposures like this one being examined by paleontologists. Photo courtesy Missouri Institute of Natural Science.

unlikely that any new, large-scale mineral production (other than existing limestone mining and quarrying) will occur in Greene County, it would not be out of the ordinary for deep drillholes to encounter mineral deposits of considerable size. However, the costs of acquiring land rights and environmental permits, plus the expense of opening new mines, would require a very large ore body in order to turn a profit.

Significant discoveries are likely, though, in cave resources. The number of caves in Greene County has grown steadily over the past half century, from barely a dozen in the 1950s to nearly 400 today. It is not unusual to encounter caves during construction projects. Perhaps the most significant such find in recent memory was Riverbluff Cave, encountered in road construction and now considered one of the most significant paleontological caves in the nation. There is no way of knowing how many other amazing caves may lie hidden, awaiting a sinkhole collapse, the intervention of construction machinery, or persistent exploration by cavers.

How Karst Works

Surprisingly, the same carbonate-based stratigraphy that hosts mineral deposits, grows caves and nourishes springs while providing a virtually unlimited supply of limestone for construction purposes, also doubles as a highly productive groundwater aquifer. In fact, the rock formations from the surface down to the basement comprise a "superaquifer" that sup-

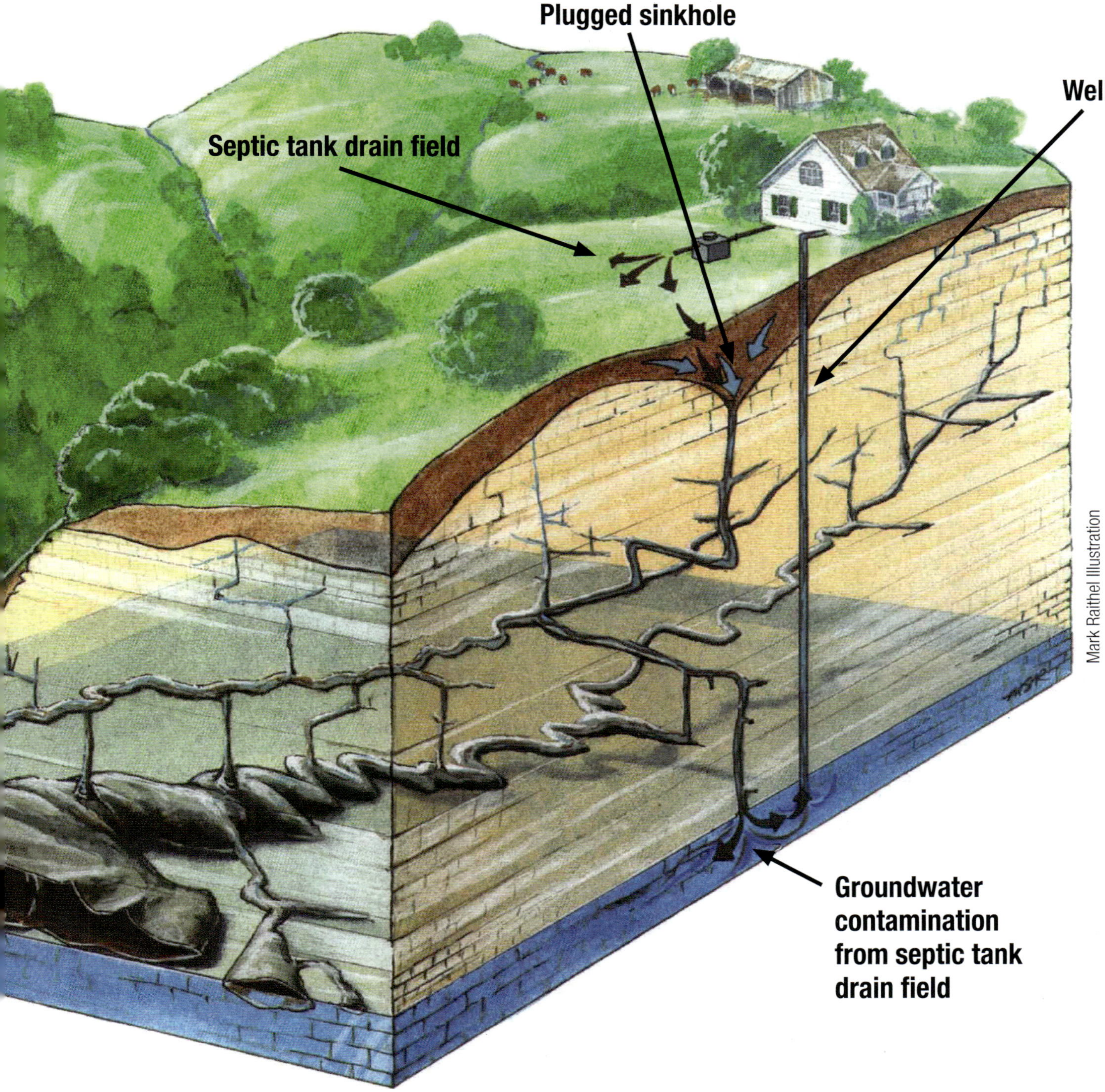

plies groundwater to a multiplicity of users who have drilled wells to various depths in the rock column.

Springfield and Greene County are geology-friendly places. The pleasant landscape, enhanced by clear streams, surging springs, and rock outcrops, offers stunning views while sustaining basic needs for raw materials and water. We have the fundamentals of positive life and work; the challenge is to manage the geologic resources in sustainable ways, for this and future generations.

CHAPTER 1

Landscape

Erosion by streams is by far the most powerful agent of landscape change in Greene County, followed by the slower, more subtle changes caused by dissolution of limestone bedrock in groundwater.

Cassville native Dr. Curtis F. Marbut mapped Missouri landscape features on horseback.

Geomorphology [literally, the study of earth forms] is the subdiscipline of geology dealing with the development of landscapes. Geomorphic study of Greene County reveals a long process of erosion by streams and dissolving away (dissolution) of primarily limestone bedrock that has produced the familiar landscape we call our home country. Surprisingly, the basic outline of Missouri's major landscape features appeared in 1896 on a map by Dr. Curtis F. Marbut, a scientist of towering accomplishments who hailed from Cassville, Missouri. That he was able to produce such a map before topographic contour maps became available and more than a century before modern satellite imaging systems came into use is testimony to the observational skills of scientists who traveled on horseback or simply walked the hills and valleys. Slow-paced, close-to-the-ground travel allows time to observe subtle landscape changes that would be missed if one were riding in a car at 70 miles per hour. The 19th century work of Dr. Marbut was followed in 1938 by N. M. Fenneman's monumental study of the landforms of the United States, in which he delineated and named Greene County's two landscape provinces, the Springfield Plateau and the Salem Plateau. These two landforms are separated by a prominent line called the Eureka Springs Escarpment. An escarpment is a difference in elevation, usually caused by the effects of weathering (erosion) on rock units. In this case, the Springfield Plateau is produced on mostly limestone bedrock, while the Salem Plateau has formed on older, mostly dolomite bedrock (dolomite is a form of limestone that has a high magnesium

About 65 percent of Greene County and essentially all of Springfield are underlain by a massively-bedded, thick limestone unit called the Burlington Limestone. The upper surface weathers to a very irregular dark gray to black color, while the underlying fresh rock is light gray with streaks of white chert. Vertical grooves are drill marks left after roadcut blasting. Hwy. 65 at Battlefield Road.

Looking west toward the face of the Eureka Springs escarpment, marking the dividing line between the Springfield Plateau and the Salem Plateau, eastern Greene County.

content) bedrock that reacts somewhat differently to erosion than does limestone. The difference in elevation of the Springfield and Salem surfaces averages about 100 feet, and is most pronounced in the vicinity of Eureka Springs, Arkansas, for which it is named.

The Eureka Springs Escarpment crosses the northeastern corner of Greene County, defining the edge of the younger (Mississippian age) rock outcrop area. The most prominent rock formation is the Burlington Limestone, named for its type section near Burlington, Iowa, so the escarpment is sometimes called the Burlington Escarpment. The bedrock formations of the Springfield Plateau—limestones, dolomites, shales and siltstones—dip (slope) gently to the north and west. On the geological map, the Eureka Springs Escarpment is a highly serrated, irregular line that is actually the receding edge of Mississippian-age bedrock that once was much more extensive to the south and east. Proof of this lies in the small, isolated remnants (called outliers) of Mississippian bedrock that are now islands surrounded by older rocks of the Salem Plateau. In the same way, there are remnants of a third erosion surface in western Greene County, in the form of outliers (monadnocks) such as Sand Mountain. These are all that is left of a younger, Pennsylvanian-age sedimentary blanket that has been almost entirely removed by erosion.

Erosion by streams is by far the most powerful agent of landscape change in Greene County, followed by the slower, more subtle changes caused by dissolution of limestone bedrock by groundwater. However, the works of humankind, using earth-moving machinery and explosives, have eclipsed natural processes in modern times. The tendency is to flatten

A prominent landmark, Fair Grove Mound is an outlier of Pennsylvanian bedrock, a remnant of a once-continuous blanket of clastic bedrock over Greene County.

high places and fill in low places, to produce a more even surface on which to build structures and install infrastructure such as roads, railroads, utility lines, and sewer services. Despite the major changes brought by "development," there are many opportunities in the City of Springfield and especially in Greene County, to see natural processes at work.

Total relief [difference between highest and lowest elevation] is 608 feet. Greene County's highest point is 1518 feet, at the top of a hill near Bassville, and the lowest point is 910 feet, on the western county line where the Sac River enters Lawrence County.

The Springfield Plateau is largely defined by the outcrop area of the Burlington-Keokuk Limestone (commonly called "the Burlington"). In the type sections near Burlington and Keokuk, Iowa, there are two rock units that have distinctive characteristics, but as one follows the outcrop areas southwestward through the Columbia, Missouri area and on to the Springfield Plateau, the formations become so similar that only skilled stratigraphers [geologists who specialize in stratified (or layered) sedimentary rocks] can tell them apart. So they are frequently brought together as a hyphenated name, Burlington-Keokuk.

Wherever the Burlington Limestone is found, it shows the effects of dissolving (dissolution) by groundwater. Chemically, limestone is calcium carbonate, which is vulnerable to acidic groundwater, which produces numerous and sometimes extensive cave systems, springs, and sinkholes. Driving around Greene County and even in Springfield, these landforms are common and distinctive features of the landscape.

Shaded Relief Map of Greene County

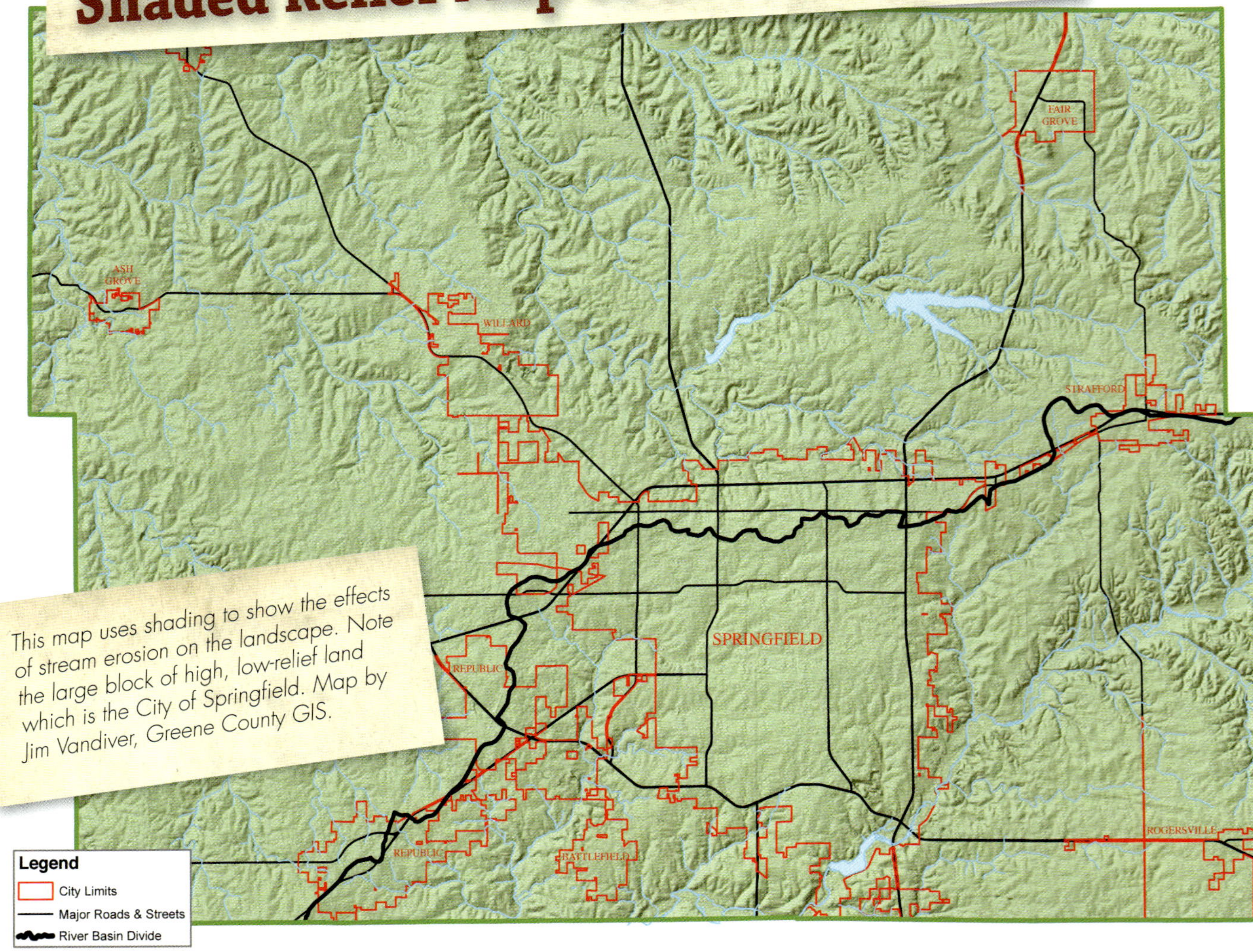

This map uses shading to show the effects of stream erosion on the landscape. Note the large block of high, low-relief land which is the City of Springfield. Map by Jim Vandiver, Greene County GIS.

If all the vegetation, soil and residuum were stripped off the Burlington Limestone in Greene County, the bedrock surface would be extremely uneven and rugged-looking. A trip down Hwy. 65 through the east part of Springfield, or along I-44 through the northern part of the city shows—in the roadcuts—what this surface might look like. In building the highways, MoDOT left the Burlington erosion surface relatively undisturbed, as a scenic feature. The effects of weathering over a long period of time have produced protruding stone monoliths called "pinnacles," separated by corresponding low features called "cutters." The presence of cutters and pinnacles–called gargoyles in this book–across the entire Burlington outcrop area makes estimating construction costs very challenging. For example, a test drilling may hit a cutter, and the engineer will conclude that there is plenty of overburden for building a basement, but another drillhole just a few feet away may hit a pinnacle, requiring costly bedrock removal.

Drainage

The Greene County landscape has been exposed to weathering and erosion for a very long period of time--long enough for a fairly complex erosional history to develop. Springfield sits astride the divide between the Missouri River Basin to the north and the White River Basin on the south. The divide runs from the eastern county line, southwesterly across the county. North of the divide lie the watersheds of the Sac, Little Sac, and Pomme de Terre rivers, and that of Turnback Creek. To the south, all of the drainage enters the James River and its tributaries, which enter the White River.

Over time, a dendritic [literally, "like a tree"] drainage pattern has developed on both sides of the divide. From the air, the stream patterns resemble the veins of a leaf, extending into all parts of the landscape. The dendritic pattern is influenced to some extent by rock structure (see section on lineaments, page 41), and by the undermining influence of sinkholes produced by groundwater dissolution.

Digging more deeply into the geology of the Springfield-Greene County landscape reveals a complexity beyond the scope of this book. For in-depth study, refer to the following sources that trace the progress of landscape geology from early concepts to current thought. Please note that the study of geology is an evolutionary process. That is, the level of understanding is based on the work of past scientists, and—generally speaking—the more study, the more complete the understanding becomes. For example, in the early part of the 20th century the sources of spring water were poorly understood. However, after the advent of water tracing using fluorescent dyes came into general use, the sources of many springs were identified, making it possible to address and overcome contamination and flow problems. In the same way, understanding how landscapes evolve through time helps address questions about the effects of development projects on the future livability of our part of the world.

Eroded surface of the Burlington Limestone, exposed in the spillway of McDaniel Dam. Fresh exposures will be light gray or tan; weathered, it becomes dark gray to almost black.

For Further Reading:

Fenneman, N. M., 1938, Physiography of the Eastern United States: McGraw Book Co., New York, 714 p. Landmark study of the nation's landscapes.

Bretz, J Harlen, 1965, Geomorphic History of the Ozarks of Missouri: Mo. Geological Survey and Water Resources, Vol. 41, 2nd. Ser., 147 p. Seminal study of the Ozark landscape.

Hawker, Jon L., 1992, Missouri Landscapes: A Tour Through Time: Mo. Dept. of Natural Resources, Div. of Geology and Land Survey, 326 p. Links landscape evolution with the ever-changing botanical and zoological world.

Marbut, Curtis F., 1896, Physical Features of Missouri, in Surface Features of Missouri: Mo. Geological Survey, Vol. 10, pp. 11-109. First delineation of major Missouri landforms.

CHAPTER 2

Footprint

We must piece together the record of geologic time from many sources, each of which records some of the historical puzzle.

Precambrian granite exposed at the surface in southeastern Missouri. Rock similar to this is found in the "basement," some 2,000 feet below Greene County's "footprint."

Greene County sits atop an incredibly long geologic history, resting on a "basement" of unimaginably old igneous rocks, upon which have accumulated, layer by layer, a series of sedimentary rocks. In geological parlance, "basement" rocks are the oldest, deepest, most different rocks beneath a landscape. In Greene County, the "basement" is some 2,000 feet—nearly a half mile—beneath the surface. The granitic rocks found there date from a time when the Earth was much younger, before there were extensive oceans and before widespread erosion and sedimentation began. They are called "Precambrian" (literally, "before the Cambrian"). Similar Precambrian rocks in eastern Missouri have been dated at 1.2 to 1.4 billion years old (Kisvarsanyi, 1974).

All we know about Greene County's basement rocks comes from a single deep hole that City Utilities drilled in 1972, through the overlying sedimentary rocks into the granitic basement. The drillhole penetrated 70 feet into Precambrian granodiorite (a granite-like rock type). We are learning a lot more about the basement rocks–and the younger strata beneath Greene County's "footprint" as well–thanks to drilling by Springfield City Utilities, at its Southwest Power Station site near Battlefield. The utility has completed a new, coal-fired electrical generating unit beside its existing unit. The new unit came on line in early in 2011.

Because of international concern that stack gases from fossil fuel-fired power plants will

(continued on page 13)

Layered (stratified) sedimentary rocks exposed in a new road cut on the Branson Airport access road.

Stratigraphy is the subdiscipline of geology that deals with layered (stratified) sedimentary rocks. Early geologists working in England noticed that rock units occurred in layers or strata, and found that they could trace these distinctive layers from place to place. So they began to map the extent and thickness of rock units, and gave them names—The Old Red Sandstone, Millstone Grit, Burlingstone Lime—and thus Stratigraphy was born.

Geologists also learned that fossils they found in strata could be used to recognize—or correlate—rocks of the same age, and thus began the sub-subdiscipline of Biostratigraphy.

Geologists who specialize in stratigraphy are called stratigraphers, and the science has evolved into a very complex field of inquiry. The first stratigraphers classified rock units according to age and gave them names that were largely taken from English place names. For example, the Devonian, from Devon; Cambrian, from Cambria, and so on. These names were retained when stratigraphy began in the U.S., but individual rock units—called formations—were named for local places where the formations are best exposed. For example, Missouri rocks of Cambrian age are well exposed in southeastern Missouri, so their names are taken from exposures in that area, as Lamotte Sandstone, from Mine la Motte; Eminence and Potosi Dolomites, from the towns of Eminence and Potosi, and so on.

The definitive stratigraphic framework in Missouri is established by the Missouri Geological Survey in Rolla, a program of the Dept. of Natural Resources. A general overview is "The Stratigraphic Succession in Missouri," 1995, by Thomas L. Thompson. Much more detail is offered by a series of reports on five major geologic systems in Missouri: Part 1—Cambrian System; Part 2—Ordovician System; Part 3—Silurian & Devonian Systems; Part 4—Mississippian System; and Part 5—Pennsylvanian System. All of these reports are available in libraries, or direct from the publisher, The Geology Store, Missouri Dept. of Natural Resources, Div. of Geology and Land Survey, P.O. Box 250, Rolla, MO 65402-9250.

Cross Section of CO_2 Injection Test Site

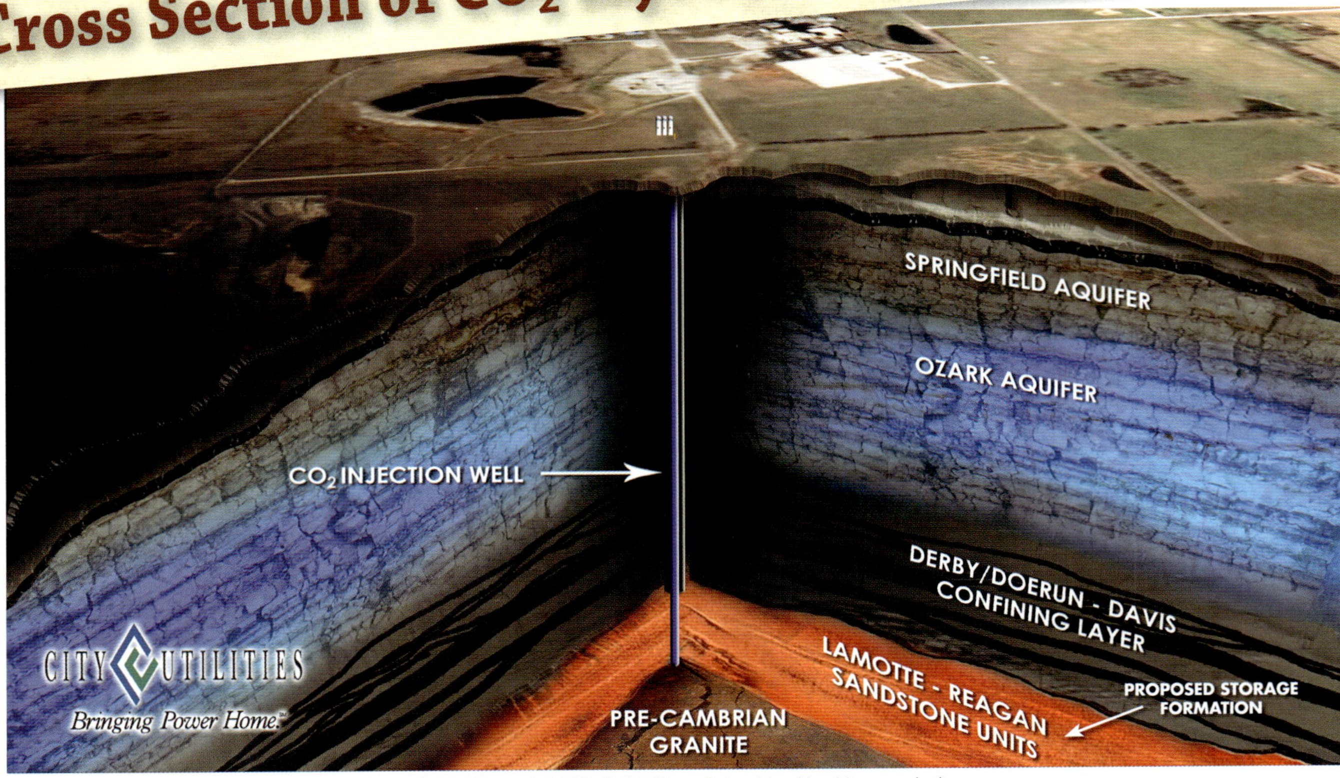

increase the rate of global warming, City Utilities brought together a multi-agency and university team to determine whether CCS (Carbon Capture and Sequestration) technology could be used in conjunction with the new power plant. The concept seems simple; capture carbon dioxide from stack gases, pump it underground through an injection well, and sequester it in an unused rock formation that is separated from groundwater aquifers by a "bulletproof" confining layer.

If the concept seems simple, the execution is a formidable technological and political challenge. We know there is a potential storage formation, and it has a confining layer, but how would the storage formation accept large amounts of carbon dioxide, and how effectively would the confining layer prevent any leakage? These and many more questions are being addressed by test drilling at the plant site. The first borehole has produced rock core into the basement, and detailed analysis is being done under a project funded by federal appropriations, City Utilities, and its partners. When the drilling is completed, an informed decision can be made on whether to pursue sequestration.

The perspective diagram above clearly shows how CCS relates to the two major groundwater aquifers beneath Springfield, and how important it is to be certain the process will work, before it is implemented. In this area, deep holes are almost always drilled in search of water supply. There are no indications that significant deposits of valuable minerals lie deep beneath the land surface, and no oil or gas is known to be present. This could change in the future, should concepts such as carbon dioxide sequestration ever become viable, or other reasons arise to justify the high costs of deep drilling.

We can only guess how long the Precambrian surface lay exposed to erosion before the first sedimentary rocks were deposited upon it, but it may have been a very long time. Granodiorite is an igneous rock, that is, one that formed from a molten mixture of mineral-forming materials. It is coarse-grained, meaning that it underwent a relatively long period of cooling, giving individual minerals time to form crystals. This history suggests that the granodiorite once had an overlying cover of finer-grained rock—perhaps lava spewed out from erupting volcanos—that was removed by erosion prior to the deposition of the first sedimentary rocks.

Erosion cannot occur absent an ocean, which powers the hydrologic cycle that produces rainfall. Rainfall causes streams to form, and streams erode the landscape. We can only guess how long erosion continued before rising sea levels submerged the ancient landscape and sediments began to be deposited.

We know that an ocean was present during Cambrian time because the rock formations resting on the Precambrian are those typical of near-shore marine environments. The oldest of these is the Lamotte Sandstone [named for exposures near the village of Mine La Motte, in Madison County], a rock that crops out at the surface in the St. Francois Mountains in eastern Missouri, but lies under approximately 2,000 feet of sedimentary rocks in southwestern Missouri. It is encountered in drillholes all across the state to Greene County and beyond. The Lamotte has structures typical of beach sand deposits in shallow, near-shore environments. Mixed with the beach sand are fragments of the eroded Precambrian rocks on which it rests. There are few fossils, partly because the organisms believed to be living at the time were relatively simple and few in number, and partly because the near-shore beach environment had a lot of wave energy that tended to destroy organisms before their hard parts could become fossilized.

Close-up of granite on Taum Sauk Mountain, St. Francois Mountains, southeast Missouri. Gray and green blotches are lichen growths.

The ancient sand deposits that became today's Lamotte Sandstone were the beginning of what is called the Paleozoic Era. This was

(continued on page 17)

Footprint of Greene County

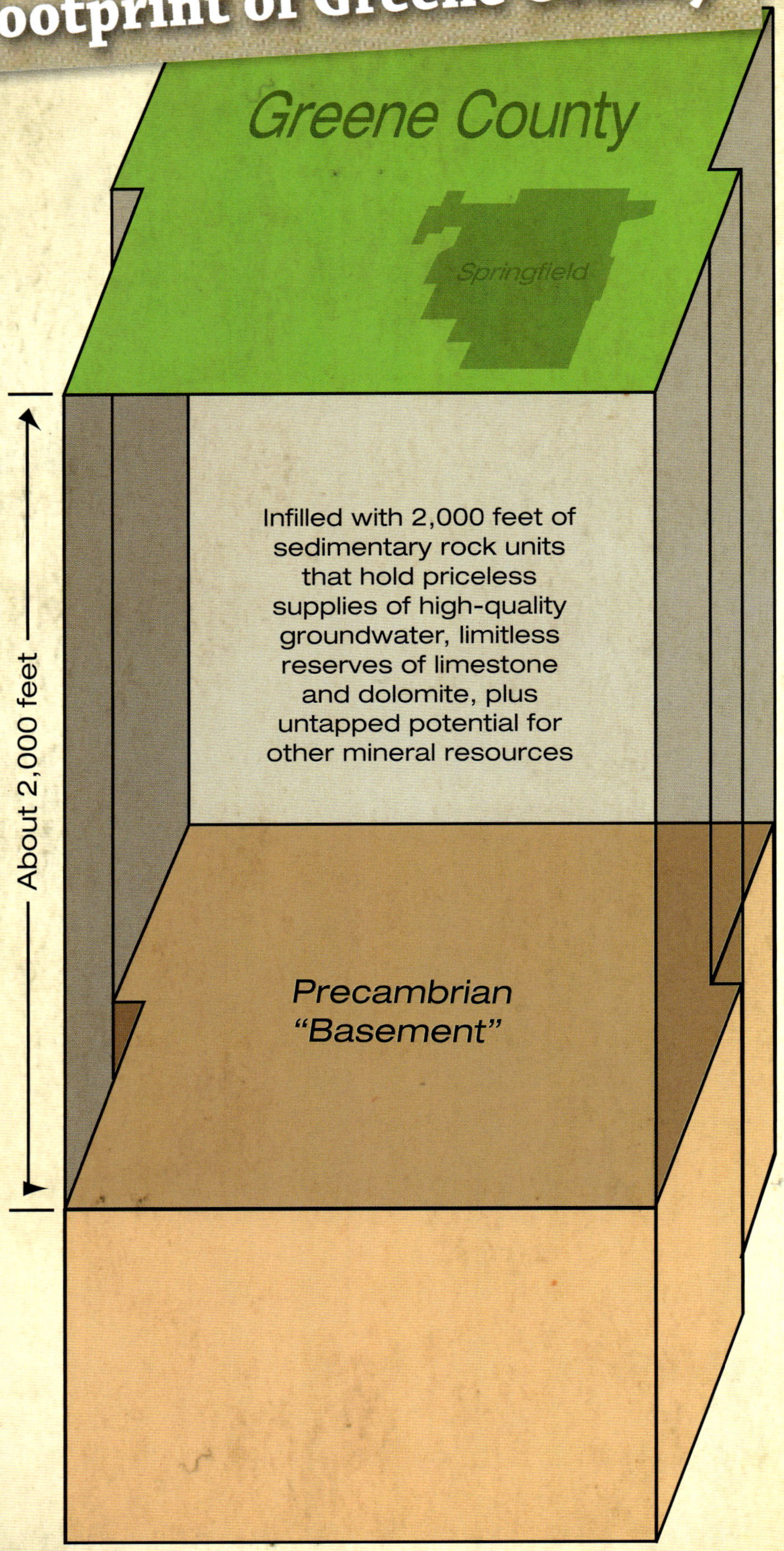

The geologic history of Greene County is contained in some 2,000 feet of sedimentary rocks that rest on the Precambrian "basement," which consists of igneous and metamorphic rocks.

Unfortunately, the record of the rocks is not continuous. "Reading" the history is like reading a book with missing pages. Geologists call a gap in the record [missing page] an unconformity.

The largest unconformity in Springfield's geologic history may be the one between the Precambrian basement and the first sedimentary rock formation, the Lamotte Sandstone. This gap ended when the young Earth developed oceans large enough to support the hydrologic cycle to generate rainfall, which in turn formed runoff streams that eroded the exposed land areas, creating sediment that eventually became rock formations.

Another major gap spans the Silurian and Devonian

Periods, including the Age of Fishes. The record of the Paleozoic Era ends with the Pennsylvanian Channel Sands.

There is no record of the Mesozoic Era under Greene County's footprint, which explains why no dinosaur fossils are found in this area. Instead of marine conditions accumulating sedimentary strata, Greene County seems to have been a land area undergoing erosion for a very long time, producing a stream-worn landscape with karst features, deep residuum, alluvial deposits, and fossils dating from the Ice Ages of the Pleistocene Epoch.

To find out what happened during the "missing pages" of geologic history beneath Greene County's "footprint," one must look elsewhere in Missouri or in other parts of the world where records of those times are preserved.

NOTE: Unconformities are TIME gaps, not PHYSICAL gaps.

Drawing not to scale.

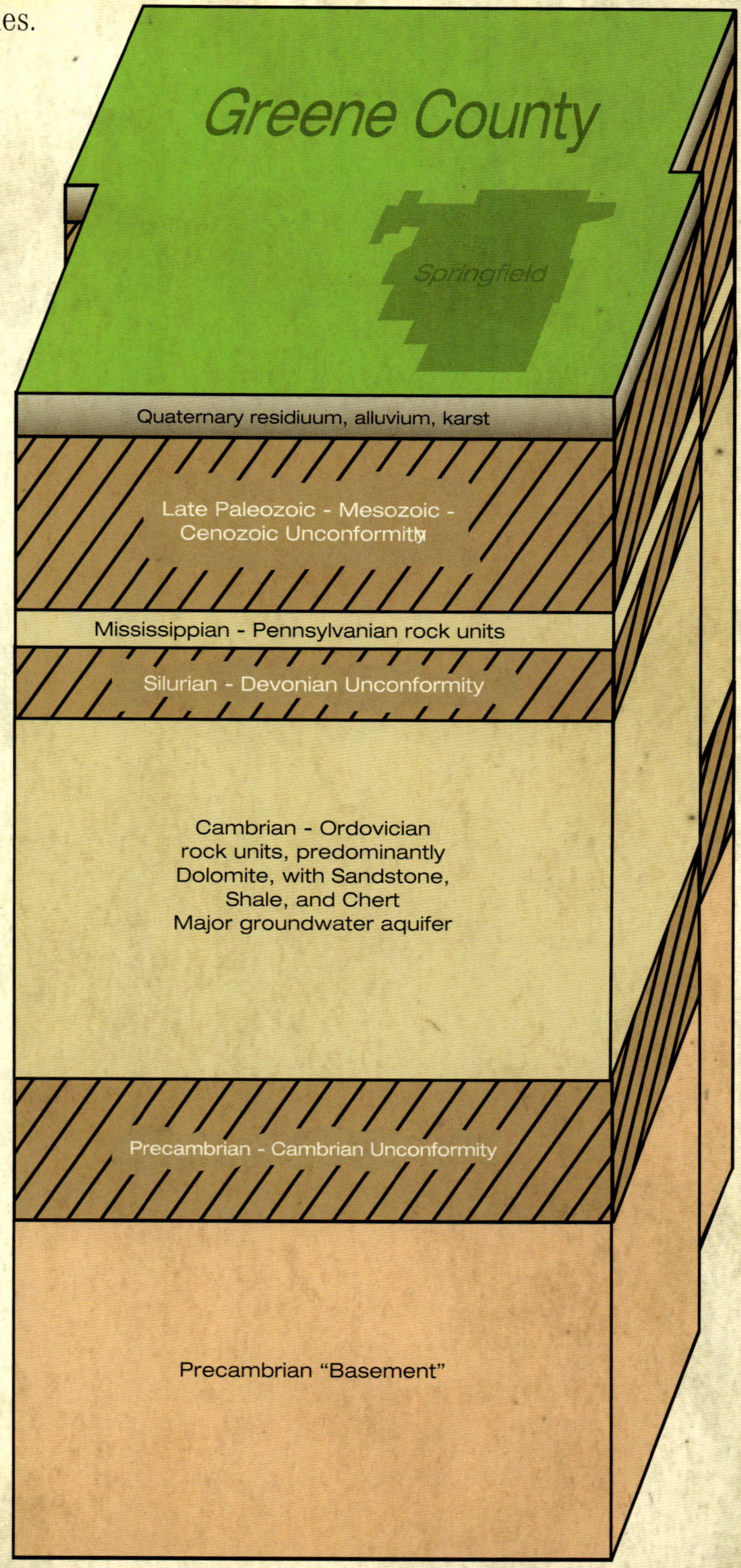

Formations

"Geologists use the term formation to designate a rock unit that has uniform characteristics, is tabular in form, and is large enough to be mappable. Formations are commonly named for a geographic feature, or area, where they are first studied. The locality where a specific formation is first described usually becomes known as the ***type section*** *or* ***type locality****."* (Unklesbay and Vineyard, 1992, p. 5).

Formation names usually begin with the name of a type section, followed by a modifier, either the word "formation" or a rock name. If a formation is almost entirely one rock type, that rock will be used as a modifier. For example, the Burlington (named for its type locality in southeastern Iowa) Limestone is a limestone unit with minor amounts of chert. On the other hand, the Roubidoux (named for its type locality along the Roubidoux River in south-central Missouri) Formation is so-named because it is a mappable unit consisting of about 80% dolomite and 20 % sandstone. When speaking of a particular rock unit, the name is considered formal and the words are capitalized, as Burlington Limestone or Roubidoux Formation. But if the topic is just the sandstone of the Roubidoux—not the entire unit—then it is rendered "Roubidoux sandstone."

a very long time—hundreds of millions of years—during which what is now southwest Missouri was a seascape, flooded by a succession of ancient oceans of varying depths, in which sediments such as sand, silt, gravel and chemical precipitates of lime were deposited. For the most part, this was continuous accumulation of sediments, which became the geologic history of the region. But the geologic record has some gaps—called unconformities—where there is no record of what was happening. For example, the Devonian Period is often called the Age of Fishes, because of the remarkable fossil record of fishes that lived in that time period. Unfortunately, there are no Devonian rocks in Greene County, so to learn what was happening during that unconformity, or gap in the record, we have to look to southeastern Missouri, or elsewhere in areas that do have Devonian rocks. In fact, there is no place on Earth which has a complete, continuous and uninterrupted record of geologic history. Instead, we must piece together the record of geologic time from many sources, each of which records some of the historical puzzle. Not surprisingly, this part of geologic inquiry is called historical geology.

The Paleozoic Era

By far the longest record of Earth history beneath Greene County's "footprint" is that of the Paleozoic [literally, "old life"] Era. Geochronology—the timing of events on Earth—places the beginning of the Paleozoic Era about 570 million years ago, when

Blue Spring on the Current River—some 300 feet deep—flows from the Eminence Dolomite. In southwest Missouri, the Eminence is deep below the surface, part of the prolific Ozark Aquifer that supplies groundwater to deep wells in Greene County.

complex forms of life first appeared in the geologic record. The Paleozoic lasted some 225 million years, or until 245 million years ago, when the Mesozoic [literally, "middle life"] Era began. The Paleozoic Era was well underway when the beach sands that became the Lamotte Sandstone began to be deposited. We have no record of the Early and Middle Cambrian; the first Missouri rocks date to the upper part of the Cambrian Period. The beach sands of the Lamotte came from erosion of what is called the Canadian Shield, lying far to the north. This vast region of igneous and metamorphic rocks is much older and higher than Missouri, so through time it has supplied sediment to build up the stratigraphic record in Missouri. There is a tendency to think that what we now call Missouri has always been where it is today—north of the equator and west of Europe—but that is not the case. Early in Earth history there was a large continental mass called Pangaea, when North and South America were joined to Europe and Africa. Looking at a globe, one can readily visualize that if you moved the continents together, they would fit. Moreover, the North Pole has moved around in the past, and there have even been polar reversals. Mountain ranges have been uplifted and eroded down, sea levels have risen and fallen, and there have been multiple Ice Ages in Earth history. For most of the Paleozoic Era, Greene County would have been under sea water of varying depths, mostly shallow-water marine conditions. We can say this because the marine fossils we find are organisms that today live in shallow seas—corals, brachiopods, crinoids and mollusks.

After the Lamotte Sandstone deposition was complete—some 180 feet of sand—sea level rose—or the land subsided. Instead of sand, lime muds were precipitated from the ancient oceans. Eventually these chemical precipitates became limestone and later underwent a process called dolomitization to

Ripple-marked Roubidoux Sandstone in an outcrop along Cabin Creek, northeast Greene County. Ripple marks, mud cracks, and even raindrop impressions are common in the Roubidoux, denoting a sea-margin environment when the sand was deposited that became sandstone bedrock.

become dolomite—calcium magnesian carbonate. This process continued long enough to accumulate 185-260 feet of a formation we know as the Bonneterre Dolomite, named after a small town in eastern Missouri where it was first described. The Bonneterre [literally, "good earth"] contains extensive reef structures that host rich sulfide ores from which lead, zinc, copper, cobalt and silver have been mined. Unfortunately, none of this mineral wealth is known in the Bonneterre strata that lie beneath Greene County. Shallower marine conditions followed the Bonneterre, producing the Davis Formation, 140-155 feet thick, consisting of near-shore shales, siltstones, sandstones, and limestones, as well as thin limestone conglomerates. The varied nature of the sediments implies an environment of shallow water with streams supplying mud and sand. Imagining ancient terranes is naturally biased because we think of present-day conditions, where rivers have names, drain well-known landscapes, and have chemical characteristics that we are familiar with. But remember that in the Cambrian Period there were no land plants, no forests, no terrestrial animals, and the water chemistry of both the oceans and the rivers reflected the contemporary atmosphere and the chemistry of the igneous rocks being eroded. There was no soil as we know it today; soil has organic components, which were absent in the Cambrian environment. Nonetheless, sediments accumulated in marine conditions, life prospered in primitive oceans, and some of the living things—especially ones with shells or other hard parts—became entombed in sediments, hundreds of millions of years later to reappear as fossils.

After deposition of the Davis Formation, water depths increased and a series of three limestones followed, which eventually became the Derby-Doerun, the Potosi, and the Eminence dolomites. The names reflect the areas in east-central Missouri where the units were first named. They lie far below the surface in Greene County, and can be traced continuously in drillholes across the state. Bear in mind that the dolomites so prominent in Missouri's geologic column apparently began as lime muds similar to those found today off the coast of Florida. During the long process of becoming rock, lime muds are compressed by the weight of overlying (younger) sediments, and go through a process called lithification [literally, becoming rock]. At some point—whether during or after lithification—the limestone becomes dolomite, or dolostone. The two rocks are chemically similar, but dolomite has a magnesium component that limestone lacks. Dolomite is prominent among the older sedimentary rocks in Missouri, but relatively rare in younger sediments.

The Derby-Doerun is dolomite interbedded with shale and siltstone. Beneath Greene County, its thickness ranges from 85-105 feet.

The Potosi Dolomite is conformable with the Derby-Doerun, but differs in having thick, massive beds of dolomite that tend to have geode-like voids called "vugs." These small openings are frequently lined with crystals. In its outcrop area, the Potosi is noted for its quartz druse (sometimes called "blossom rock"), and for its extensive barite mineralization. Beneath Greene County, it is primarily a groundwater aquifer (an aquifer is a rock formation capable of storing and releasing groundwater).

Marine conditions during deposition of the Potosi Dolomite changed little as the Eminence Dolomite accumulated. In fact, the two rocks are so similar that it is difficult to tell them apart. The Eminence marks a notable increase in chert content, noticed first in the Potosi, reflecting an increase in variety of marine species. Beneath Greene County the Eminence ranges from 260-350 feet, and is known as a prolific aquifer, because groundwater readily moves through the formation. To the east, in its primary outcrop area, the Eminence is host to some of Missouri's largest and deepest springs, as well as some of the state's most significant caves.

The transition from Cambrian to Ordovician time records continuous accumulation of marine sediments, marked by shallowing seas. The Gunter Member of the Gasconade Dolomite is sandstone and sandy dolomite, with shale lenses, 20 to 55 feet thick. The Gunter is very persistent and recognizable over a wide area, so it is useful as a marker unit interrupting a thick series of carbonate rock formations that are otherwise so similar it is difficult to tell them apart. After the Gunter was deposited, water depths seem to have increased, because the rest of the Gasconade is massively-bedded, medium-grained dolomite with extensive and persistent chert layers. These cherts are thought to be algal reefs several feet in thickness,

Close-up of Jefferson City-Cotter Dolomite. Fresh rock is tan to yellowish in color; weathers gray.

Jefferson City-Cotter Dolomite in roadcut along Hwy. 65, Branson. In all except northeast Greene County, this formation is in the subsurface, part of the Ozark aquifer.

persistent over wide areas and useful for geologic mapping. They are called "Cryptozoan" reefs, meaning "secret life." From 250-450 feet thick, the Gasconade—named for its type section in bluffs along the Gasconade River—is the host rock for many of Missouri's best-known caves and springs.

Jointed and fractured Roubidoux sandstone in outcrop along Cabin Creek, NE Greene Co. Recognized as a prolific aquifer, the Roubidoux is the target for many private water wells.

Rock Formations That Crop Out In Greene County

Roubidoux Formation

With the Gasconade in place, sea levels fluctuated during deposition of the Roubidoux Formation, named for outcrops along the Roubidoux River in south-central Missouri. The formation has three prominent sandstone units separated by dolomite beds that resemble the dolomite of the Gasconade Dolomite. The Roubidoux crops out in the extreme northwestern part of Greene County, along the Pomme de Terre River and tributary creeks. In the rest of the county, the Roubidoux is in the subsurface, where it is known as a strong groundwater aquifer. The depositional environment is predominantly near-shore marine, but extensive ripple-marked and mudcracked beds record a landscape barely above sea level. Bedding-plane surfaces frequently preserve fragile features in exquisite detail, such as raindrop impressions and the aforementioned ripple marks and mudcracks. Thin beds of Roubidoux sandstone are quarried for building stone in some areas. Traditionally the stone is used for flagstones and in building construction. In earlier times, stone laid on edge produced a "patchwork quilt" effect sometimes called "giraffe rock" because of its resemblance to a giraffe's coat. The same stone, cut into uniformly narrow pieces, produces a brick-like effect in building construction. While there are no Roubidoux stone quarries in Greene County, there are numerous buildings faced with the stone. A notable example is the Roaring River Lodge, at Roaring River State Park near Cassville.

Jefferson City-Cotter Dolomite

After deposition of the Roubidoux Formation, there followed a lengthy time of shallow marine conditions during which the monolithic Jefferson City and Cotter Dolomites were laid down. These two rock units are so much alike that it is very difficult to tell them apart in field exposures, so the names are frequently hyphenated and called the Jefferson City-Cotter Dolomite. The older of the units is named for bluffs along the Missouri River at Jefferson City, while the Cotter is named for exposures near Cotter, Arkansas. The sedimentary rock likely first became limestone, then changed to dolomite over a long time span. The dolomite of the Jefferson City-Cotter is more fine-grained than the older dolomites of Cambrian age, such as the Potosi, Eminence, and Gasconade formations. Some beds are so dense and fine-grained that the stone is called "cotton rock."

The Jefferson City-Cotter was once a popular building stone. One prominent layer is called the "quarry ledge," which forms a useful marker unit for geologic mapping. Fossils are relatively rare, occurring most frequently in the chert beds in the dolomite. The Jefferson City-Cotter Dolomite may be seen in roadcuts along Hwy. 65 in the Branson area, especially at Saddlebrook and at the intersection of Hwy. 65 and the Ozark Mountain Highroad. In relatively new roadcuts, the rock is characteristically tan or yellowish-tan, but as the exposed rock weathers it turns gray. Outcrops of the Jefferson City-Cotter are found only in the northeastern part of Greene County.

Silurian-Devonian Unconformity

Deposition of the Jefferson City-Cotter dolomites was followed by a long unconformity (gap in the geologic record) that spanned two entire geologic periods—the Silurian and the Devonian—plus a considerable part of the Mississippian Period. While there is a wealth of information on what happened during this time span in other parts of the world, the only record in Springfield/Greene County comes from a few drillholes in the southwestern part of the county, where drilling passed through sections of the Chattanooga Shale and the Bachelor Formation. Geologists reason that this area was a land mass undergoing erosion, but we cannot determine precisely how much erosion actually took place, or whether there may once have been Silurian and Devonian deposits, that were eroded away, along with however much pre-Silurian rocks may have disappeared.

A rough estimate of how much time is represented by the Silurian-Devonian unconformity is about 78 million years. This time span includes what has been called the Age of Fishes.

Mississippian System-Bachelor Formation

Rock units assigned to the Mississippian System are by far the most prominent exposed rock formations in Greene County. The sequence begins with the oldest unit—the Bachelor Formation—which lies unconformably on the Jefferson City-Cotter Dolomite. The Bachelor is a poorly-sorted, pale green quartz sandstone or conglomeratic sandstone which is overlain in some areas by green, sandy shale. The sandstone is composed of particles from silt-size to pebbles of chert derived from the underlying dolomite. The grains are angular to sub-rounded. The green

Roadcut along Hwy. 13 north, exposing colorful Northview-Pierson strata.

Freshly-exposed Burlington Limestone—shown here in a roadcut at Glenstone and James River Expressway—is brown on the upper surface and light gray on solid rock. The stone soon weathers to shades of gray.

color results from disseminated grains of a mineral called glauconite. Rounded, black phosphate nodules ranging from 1/8-th inch to 1 inch in diameter are also present. The sandstone weathers tan to brown as a result of iron oxidizing in the glauconite. The cementing material is siliceous or calcareous.

The Bachelor Formation is from 2 inches to 2 feet in thickness. Where the shale unit occurs, it comprises the upper half of the formation. Where the unit is exposed, the upper and lower contacts are hard to find. Generally, the unit is overlain conformably by the Compton Limestone and overlies the Cotter Dolomite unconformably.

Compton Limestone

The Compton Limestone consists of a light to dark greenish to bluish gray, finely crystalline limestone which becomes dolomitic in some areas. The limestone is thin to medium-bedded with the beds separated by thin green shale layers ½ to 1 inch apart. Weathering of the limestone with its shale layering imparts a highly nodular, thin-bedded appearance on even the medium to thick beds. This weathering pattern is characteristic of the unit. The dolomite is light brown to mottled pink. The argillaceous matter in the dolomite is concentrated in greenish-gray globs rather than as partings. Outcrops of dolomite are massive and the weathered surfaces are gray, tan, or brown. Weathered surfaces show an abundance of small crinoid columnals in relief and rust spots where the included pyrite has weathered. Pyrite is common in disseminated crystals and vug fillings (vugs average an inch in diameter). As they weather, areas surrounding the vugs turn brown from the development of limonite. Glauconite is common in the rock. The Compton is exposed in the north and west parts of the county.

The Compton overlies the Bachelor formation with no evidence of an erosional break. In most exposures, the Bachelor is too thin to be observed or is not exposed. The overlying Northview contact is relatively sharp and easy to recognize. Regional evidence indicates that the Northview is conformable with the Compton.

Pierson Limestone

The Pierson Limestone consists of a silty dolomite to dolomitic cherty limestone. In the north part of the county, the Pierson is composed of three units. The lowermost unit is a massive, finely crystalline, silty dolomite and dolomitic limestone. Weathering produces brown to tan colors and rounded surfaces. Weathering of the commonly found calcite-filled vugs creates a pitted appearance. The middle unit is likewise tan to brown, medium-bedded dolomitic limestone with dolomites and layers of lenticular chert. The chert ranges from gray to blue with tan to brown exteriors and occurs as nodules, lenses or discontinuous beds. A thin to medium-bedded gray, crinoidal limestone containing some brachiopods makes up the upper unit. This weathers into a thin-bedded appearance. To the south, the units are not as well defined.

Fellows (1970) described the Pierson at Turner Station as "*...a gray to brown dolomitic limestone 35 to 42 feet thick. The limestone has a fine-grained matrix with abundant fossil fragments, mainly stem segments of crinoids. Color changes from gray to brown on weathering and the weathered surfaces are characteristically pitted.*"

Elsey Formation along east Sunshine near Turners Station. The formation is approximately half limestone (gray) and half chert (light-colored with fractures).

Throughout the county, the Pierson is mantled with residuum from the Elsey and the Burlington-Keokuk formations. Where steep slopes occur, the Pierson may crop out in bluffs. Generally, it forms a steep slope between the shallower slopes of the underlying Northview shales and the overlying Elsey limestone and chert. Some of the best exposures are found in roadcuts along Hwy. 13 to the north and Hwy. D east of Springfield.

Elsey Formation

The Elsey Formation is a dense, light gray limestone containing 25 to 50 percent chert. The limestone is finely crystalline to micritic and tends to break with a conchoidal fracture. Fossils are sparse, but some crinoids have been found in the weathered, finely crystalline limestone in the basal 6 to 8 feet of the formation. The fossils show on the weathered surfaces. Weathering produces rounded beds 6 inches to 2 feet thick. The Elsey chert occurs in nodules or long lenses and beds from 6 inches to 1 foot thick in the formation. The chert has smooth or gently rounded surfaces or may be irregular and knobby. The

Residual chert and red clay from the Warner formation, western Greene County.

distinctive cherts are white to gray and have irregular brown mottling with some brown spots from 1/16th inch to ¼ inch in diameter. Where the limestone is in contact with the chert, the limestone dissolves readily. The cherts tend to fracture easily and weather out of the formation, leaving hollow places in outcrops.

The cherty Elsey residuum can be identified by its distinctive coloration. The cherts also shatter readily when hit with a hammer and break with a right-angled, closely-spaced pattern which creates an angular, cobble-sized residual rubble. Land areas underlain by the Elsey reveal distinctive, chert-covered hillsides.

In Greene County the Elsey Formation, which ranges from 50 to 60 feet thick, overlies the Pierson Limestone conformably with a relatively sharp contact. The unit may actually be a facies of the lower Burlington and as such it merges into the Lower Burlington to the north with the loss of chert and increasing coarseness of the limestone. Outcrops of the Elsey can be found along the Eureka Springs/Burlington escarpment on the east and north sides of the county. Some outcrops also occur near Fellows Lake, along the James and Sac Rivers, in the western part of the county, and near Ritter Springs, northwest of Springfield.

Burlington-Keokuk Limestone

The uppermost Burlington-Keokuk Limestone (Keokuk Limestone) is a light gray, finely crystalline, crinoidal limestone. More or less continuous beds are usually from 6 to 9 inches thick and range from 2 to 8 feet thick, generally separated by cherty zones in the lower part of the unit. The Short Creek is a thin, persistent bed of light gray, oolitic limestone 2 to 8 feet thick at the top of the Burlington-Keokuk Limestone. It has been used as a datum for the top of the Osagean Series.

The lower part of the unit (Burlington Limestone) is a light gray to white, coarsely to medium-crystalline, crinoidal limestone. The upper two-thirds of this part is composed of pure gray to white, coarsely crystalline, highly crinoidal medium to very thick-bedded limestone with common stylolites. Chert occurs in elliptical white nodules, usually not more than 6 to 8 inches thick arranged in non-continuous layers. The lower third of the unit is composed of alternating beds of brown and gray, fine to coarse-grained, crinoidal limestone and numerous brown dolomite beds. The limestone is somewhat silty and locally siltstone beds occur at the base of the formation. The upper part of the unit is marked by a distinctly nodular, dark blue, algae-like limestone bed. Chert is rare. Locally, the Burlington-Keokuk limestones become dolomitized to a dark brown dolomite. This can be observed along the Little Sac River near Fantastic Caverns.

Burlington-Keokuk cherts are gray to white and occur in the form of either nodules or beds. They weather white or brown as boulders and cobbles which become reddish brown with intense weathering. Internal molds of fossils, quite common in the chert, tend to break free intact, in contrast with Elsey cherts found lower in the section. They are also much more crinoidal and break in an irregular manner, much more so than do the Elsey cherts. Weathered cherts from the Burlington-Keokuk are generally rounded and vary in size from small cobbles to large boulders.

Weathering of the formation causes the limestone to become rough, irregular and broken. Surfaces show crinoid columnals and other fossils in relief so that they can be removed easily. Further deep weathering creates features described by Fellows (1965) as cutters and pinnacles, a highly irregular interface between the soil horizon and the subsurface bedrock. The Burlington-Keokuk Limestone is extremely susceptible to solution and a karst topography has been superimposed on it. Numerous sinkholes are present in the upland surface underlain by the formation. Exposures of the unit cover about 65 percent of the surface area of the county. The majority of the sinkholes in the unit are shallow, saucer or bowl-shaped depressions. Some are as much as 40 to 50 feet deep. Many caves in the county are also associated with this formation. Topography develops as gentle slopes and rolling hills. In some valleys stream undercutting has developed bluffs.

The Burlington-Keokuk Limestone overlies the Elsey Formation conformably. In some areas, the Elsey probably grades into the Burlington-Keokuk, as may be seen in northern Greene County and southern Polk County. In other areas the contact is so sharp that you can put a pick point on it. Here the contact between the limestone and chert of the Elsey make a marked contrast to the coarsely-crystalline Burlington-Keokuk Limestone.

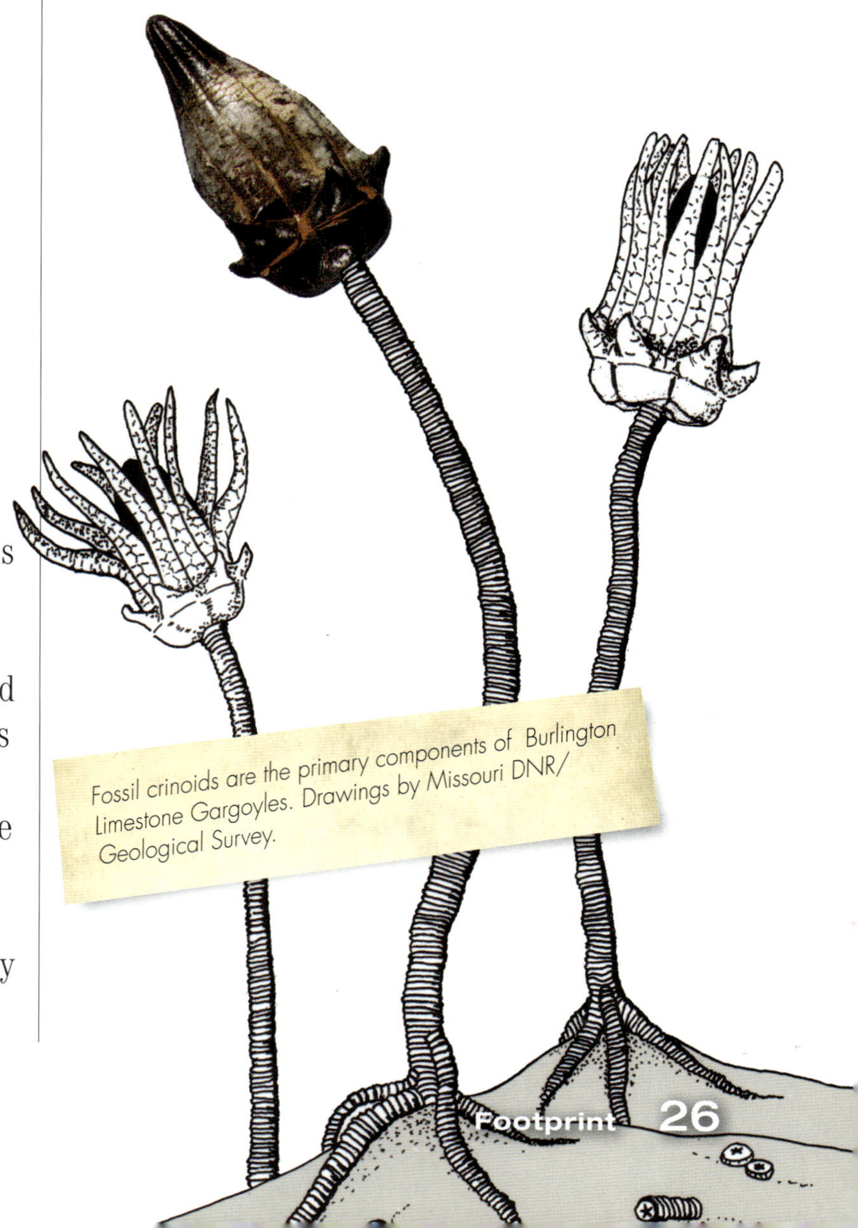

Fossil crinoids are the primary components of Burlington Limestone Gargoyles. Drawings by Missouri DNR/ Geological Survey.

Mississippian System – Meramecian Series

Warsaw Formation

The Warsaw Formation consists of medium-bedded, fine to coarsely crystalline limestone, in part crinoidal. Frequent shale partings are common. The limestone is bluish gray to light brown. Chert nodules in the formation are similar to those in underlying units.

Thickness of the unit could not be determined because of erosion, but more than 50 feet of Warsaw remains exposed in the county.

Because outcrops are similar to those of the underlying Burlington-Keokuk, it is difficult to make the distinction in a single outcrop. The Warsaw is conformable on the Burlington-Keokuk. Without the Short Creek Oolite marker bed at the top of the Keokuk it would be almost impossible to determine the contact.

Exposures of the Warsaw are confined to three areas in western Greene County. They are all preserved due to downfolding of the unit in structural basins. Exposures can be found in southwestern Springfield, southeast of Willard and south of Walnut Grove.

Pennsylvanian System—Desmoinesian Series

Channel Sandstones

Pennsylvanian Channel Sandstones consists of sandstone, conglomeratic sandstone and conglomerates. Fine to coarse-grained sandstone is thin-bedded to massive, friable to quartzose, and is red, brown, light brown, or white. Quartz grains with interstitial muscovite and chlorite are held together with ferruginous to siliceous cement. Ferruginous cement makes the sandstone red to brown; where it is lacking, the sandstone is lighter in color. Pebble to boulder-size particles in a siliceous to ferruginous sandstone matrix comprise the conglomerates. The sand and a few of the boulders are quartz; chert derived from Ordovician to Mississippian sediments predominates in the large particle size range. Fragments are angular to subrounded with some showing evidence of being reworked. Others are angular and may be abraded residuum.

Channel Sandstone occurs as resistant cap rocks, ridges and mounds and as valley filling of old stream channels in the Mississippian sediments. Channelways can be traced in the western part of Greene County, where most of the sandstone outcrops occur. Sandstone channels have been found on an intensively eroded surface of Mississippian sediments. The unconformable unit can be attributed to stream deposition in a fluvial environment.

Warner sandstone conglomerate, of Pennsylvanian age, along I-44 in western Greene County.

The Great Mesozoic Unconformity

Geologic time can be divided into four major periods: the Precambrian [literally, before the Cambrian], the Paleozoic ["ancient life"], the Mesozoic ["middle life," or the Age of Reptiles], and the Cenozoic ["recent life," or the Age of Mammals]. The first two divisions are well represented—though with some gaps (unconformities)—beneath Greene County's "footprint." But the Mesozoic Era is totally missing from the rock record. This is a huge span of time; the great American West with its lofty mountain ranges came into being. Dinosaurs arrived on the scene, developed an amazing diversity as reptilian life forms dominated the Earth's land masses, and then vanished after a giant meteorite smashed into the Earth to end the Mesozoic Era. The ordering of the continents via the mechanism of plate tectonics slowly brought into perspective the familiar geography of today's world. North and South America moved away from Eurasia and Africa, leaving the Atlantic Ocean with its mid-Atlantic Ridge in its wake. World-wide changes occurred in the Mesozoic, but in Greene County the record is simply missing. There is little in the rock record to document what happened since the scattered, discontinuous Channel Sandstones were deposited in the Pennsylvanian Period. It may be that future detailed studies may shed more light on what happened in Greene County during Mesozoic and Cenozoic eras as well.

Quaternary System—Late Pleistocene To Holocene Series Cenozoic Era

The early part of the Cenozoic Era, including the Paleocene, Eocene, Oligocene, Miocene and Pliocene are all unrepresented in the rock record. Only the Pleistocene [frequently associated with the Ice Age] and the Holocene, as represented in surficial materials, soils, and fossils, bear witness to more recent geological change.

Residuum

From all indications, Greene County has been a land area undergoing weathering and erosion since the Pennsylvanian Period. When weathering—the natural decay of rock units exposed to the elements—exceeds erosion by streams and dissolution by groundwater—a residue of insoluble materials called residuum accumulates on the bedrock. Limestone and dolomite rock units are especially prone to developing residuum, and in Greene County residuum is a familiar—if inconvenient—presence. The Burlington-Keokuk Limestone—present over some 65 percent of the county—develops a characteristic deep-red clayey residuum laced with pebble-to-boulder size chert fragments. This residuum is a familiar sight in almost all construction projects. Residuum from dolomite is usually browner in color and less cherty. Thickness of residuum is highly variable and can range from 10 to 20 feet or more.

Loess

Climate change during the Pleistocene caused world-wide glaciation that is commonly called the Ice Age. Ice sheets accumulated in polar regions and in high-altitude mountain ranges, and moved southward (in the northern hemisphere) in a number of advances, separated by warmer periods when the ice melted back during interglacial stages. Advancing glaciers operate much like gigantic bulldozers, grinding the underlying rocks into fragments, which then become frozen into the ice. When glaciers melt, all of this ground-up material accumulates, dries out, and the smaller particles get blown around by winds. This process produces a material called loess, which is composed of very small, angular grains of rock. The angularity of the grains gives wind-laid deposits the unusual ability to stand in vertical cuts, even though it appears to be highly erodible.

Glaciers never reached Greene County, even at the height of glaciation, but wind-deposited loess from melting glaciers in north Missouri probably once covered most of Greene County. Stream erosion removed much of it, but some loess can still be found in deposits of 6 inches to a foot or more in thickness. This unit is not shown on the geologic map.

Colluvium

Colluvium is a mixture of residuum and loess which accumulates on hillslopes, where the unconsolidated material migrates downslope by the action of running water and gravity. The thickest colluvium is therefore at the bottom of hills and the thinnest, near the tops. This unit is not shown on the geologic map.

Alluvium

Alluvium is sediment that has been transported by streams and deposited in stream valleys either temporarily or more permanently on river floodplains. Fellows (1970) identified two types of alluvium in the Galloway quadrangle, which also occur throughout the county. Rounded to angular chert gravel mixed with smaller amounts of red clay found in the river floodplains, shifted and changed by present-day stream action, is the first type. Sometimes limestone or dolomite fragments derived from rapid erosion of the underlying sedimentary rock formations in the area also are included. The proportion of rock fragments to clay is variable. In some areas gravel makes up most of the alluvium and in others it is nearly pure clay with little gravel.
The second type is called a river terrace, which is actually older alluvium that marks a higher-level floodplain eroded by existing streams. Remnants of previous floodplains show up as terraces or benches along the sides of the present floodplain. These terraces contain chert gravel that is well rounded and stained brown by iron oxide. Several of these terrace gravels occur throughout the county.

Alluvium deposits can be found in any stream valley. Alluvium varies in thickness up to as much as 20 to 40 feet. Floodplains typically develop rich soil zones, and terraces—being relatively higher in elevation—provide sites for farm buildings and/or residences that are relatively less prone to flooding.

Soils

Soils are derived in part from the surficial materials—residuum, colluvium, alluvium—that develop over bedrock. They have both inorganic and organic components, so they are different from strictly geologic units in that soils support plant growth. Soil science is a robust discipline that involves the identification, description, and mapping of individual soil units. There is a complex hierarchy of soil types and characteristics and a well-developed support network including the U.S. Dept. of Agriculture, Natural Resource Conservation Service, the Missouri Dept. of Agriculture, and the Univ. of Mo. Extension Service. And the Missouri Dept. of Natural Resources, Soil and Water Conservation Program deals with soil issues on a statewide basis.

Northview shale in roadcut, Hwy. 65 N. The Northview crops out in northeastern Greene County, and is encountered in wells elsewhere in the county, separating the shallow aquifer from the deeper Ozark aquifer.

Perspective

Ireland is often fondly thought of as "The Auld Sod," and we think of it as a place rich in history, with a hoary sort of ambience. By contrast, Greene County feels somehow younger, with perhaps a more adventuresome spirit. But the "footprints" of both Ireland and Greene County bear the imprint of eons of refinement of this round, rocky planet we call Earth. The time span is the same, but the events that shaped the two regions were vastly different. The same point could be made for any other place on earth.

The people of Greene County—like the Irish—are influenced by the special qualities of the "sod" they live on. It may be subtle, but it is there, and it is something to cherish, to take pride in, to understand, and to pass on—undiminished—to those who will populate the future.

CHAPTER 3

Surficial Materials

Alluvial deposit over bedrock streambed, South Dry Sac River at Lost Hill Park. Alluvium is highly variable in thickness and sediment size, including clay, silt, sand, gravel and cobbles. The upper part is the soil zone. In dry weather, the streamflow is lost to underground channels, giving rise to the name, Dry Sac.

Alluvium

What to see: Sediment transported by streams and deposited in stream valleys is called "alluvium." It ranges from clay and silt through sand, gravel, cobbles and boulders, depending on the energy level of depositing streams.

Where to see it: Any stream will have alluvium exposures. Along major streams, such as the James River, look for cut banks where the river has exposed the sediments that underlie its floodplain. On smaller creeks and branches, just look for places where erosion has exposed unconsolidated sediments.

Details: Alluvium and streams are inseparable, but the distinction between alluvium, colluvium, and residuum can easily be blurred, and more appropriately treated as soil science than as geology.

Colluvium

What to see: This material may be a mixture of residuum and loess, addressed more by soil science than geology.

Where to see it: Look for deposits at the bases of hill slopes, where it characteristically accumulates.

Details: Not shown on the geologic map, colluvium is of concern mainly in construction projects, and distinguishing it from other surficial materials may be difficult. Alluvium is moved by water; colluvium moves downslope by gravity; loess is wind-laid sediment, and residuum forms in place.

Loess

What to see: During the Ice Ages, massive continental ice sheets covered the northern half of Missouri. When the ice melted, large amounts of rock powder were left behind.

Winds picked up the dusty material and redeposited it as a blanket of material called loess. Greene County was once covered by a thin veneer of loess.

Where to see it: Loess is more likely to be found on uplands, but it may be exceedingly difficult to distinguish from normal soil materials. Exposures revealed by highway or building excavations may include loess sections.

Details: As with alluvium and colluvium, loess is more appropriately addressed through soil science than geology, in southwest Missouri.

Residuum

What to see: The unusual word 'residuum' describes the insoluble residues left behind after soluble limestone and/or dolomite are dissolved by long-term weathering. In the Springfield area, it is mostly associated with the Burlington and Elsey formations.

Where to see it: Look for residuum on the uplands, never in stream valleys filled with alluvium. Roadcuts along streets and highways, and excavations for construction projects are the best places to look. Residuum rarely forms outcrops because it is essentially loose clay and chert. Left exposed, vegetation quickly takes over.

Details: Residuum is very colorful, and may be a good place to look for fossils. Fresh residuum of the Burlington Limestone will be dark red to brown in color, with abundant white to reddish brown chert fragments. Fossils are common in the chert pieces. Thick residuum sections are frequently exposed in road construction, such as the 2008 upgrading of Hwy. 65 north of Springfield, to four lanes. During the active construction phase, residuum will appear as dark red slopes with abundant white-to-brown chert fragments. It erodes easily during thunderstorms and forms quagmires when wet, so highway builders customarily spray on a mixture of grass seed, fertilizer, and straw to establish a quick vegetative cover. Residuum of the dolomite strata in northeastern Greene County is browner in color, and the chert fragments are duller in color than that of the Burlington and Elsey formations.

New roadcut in Burlington residuum, Hwy. 65 north. Red clay with abundant chert. Commonly residuum exposures are sprayed with a mixture of seed, fertilizer, and a binder, for quick revegetation to resist erosion.

Gravel Bars

What to see: Gravel bars—familiar to canoeists who use them for campsites and/or places to take breaks from paddling—are found on most Greene County streams. They make up an interesting subset of surficial materials and provide an endless supply of rocks for children to collect.

Where to see them: Gravel bars (and their "cousins," sand bars) are common along just about every Greene County stream. The best ones are usually on the larger creeks and rivers, but they are likely to be on private property or on public lands where collecting may be forbidden, so always ask permission.

Details: "Gravel" as used by geologists is a size term, fourth in a progression—from small to large—that runs clay, silt, sand, gravel, pebbles, cobbles, and boulders. However, the common use of the term gravel refers to accumulations of stream-washed rock fragments that may range in size from sand to cobbles and even boulders. Ideally, streams do a good job of sorting sediments by size, but Ozark streams—prone to flash flooding—are more likely to produce a mixture of fragment sizes.

The ultimate source of gravel is the relatively insoluble parts—usually chert—of rock formations. For example, the Elsey Formation is roughly 50 percent limestone and 50 percent chert. When it weathers, the limestone dissolves away, leaving some insoluble clay, silt and sand, and the chert breaks into fragments that become gravel. Once freed from the limestone, chert fragments are tumbled in streams—especially in flood flows—and gradually become smoothed and rounded. On a good gravel bar, a child can discover in 10 minutes enough odd-looking rocks to confound most geologists. Fossils are commonly found, along with an almost unlimited variety of chert in various colors, shapes, and sizes. Many will have unusual markings, making simple questions like, "Mr. Geologist, what causes these funny lines in this rock I found?" a minefield to answer, eliciting responses of, "Didn't I just hear your Mother calling you?"

The best gravel bars for collecting are those along larger streams that drain areas

Gravel bars in Ozark streams have an amazing variety of rocks, varying widely in color, rock type, markings, and degree of rounding. Every flood rearranges the gravel and offers a new variety of interesting stones.

Ozark oddrocks–any gravel bar contains an amazing variety of odd-shaped, strangely-marked, and multicolored rocks. This selection includes smooth gray limestone, various cherts, crinoid fossils, and even a ringer (the black and white "rock" in the upper right is actually a piece of roadway asphalt, tumbled and rounded by stream action).

underlain by several rock formations. That produces a greater variety for collecting, and the best part is, every flood rearranges the gravel and produces a fresh collecting field. Collecting is usually best in the spring, because spring flooding tends to clean the gravel of last year's algae growth.

CHAPTER 4

Geologic Map

Geologic maps are almost always works in progress, subject to revision as newer, more detailed information becomes available.

A geologic map shows the rock units—called formations—that would be found at the surface if all vegetation, soil, and non-rock materials were stripped away, leaving nothing but bedrock. Hand-in-glove with the geologic map goes the geologic column, which is a graphic representation of the rock units arranged in a column, in order of age. The oldest rocks are always at the bottom, youngest at the top. The different kinds of rocks—as limestone, sandstone, shale, etc.—are shown by universal symbols for easy reading.

Geologic maps are essential for many reasons, but especially for construction projects of all kinds, mining and quarrying activities, groundwater studies, and city-county planning.

The process for making a geologic map is straightforward, but it requires professional training and much field work. Beginning with a standard topographic map as a base, the geologist goes to the field to find and mark on the base map all outcrops, noting the type and attitude (whether horizontal or dipping [at an angle]) of rock strata, and the elevation. The more outcrops one can find, the more complete and accurate the finished map is likely to be. In recording outcrops and elevations, the geologist may note anomalies in elevations, zones of disturbed rocks, or drainage anomalies that suggest faulting (fractures in bedrock along which there has been movement—up, down, laterally, or both). Such evidence is duly entered and refined in later stages of map construction. When field work is over, the geologist has an outcrop map (see sidebar on outcrops) that will be the framework for his geologic map.

With field work complete, mapmaking shifts to the office. Well logs—records of rock types encountered in drilling wells—include useful information such as the names of rock formations and their elevations. So the geologist plots this information on his outcrop map, to supplement what is already there. With all of the basic information now shown on the working map, the geologist uses

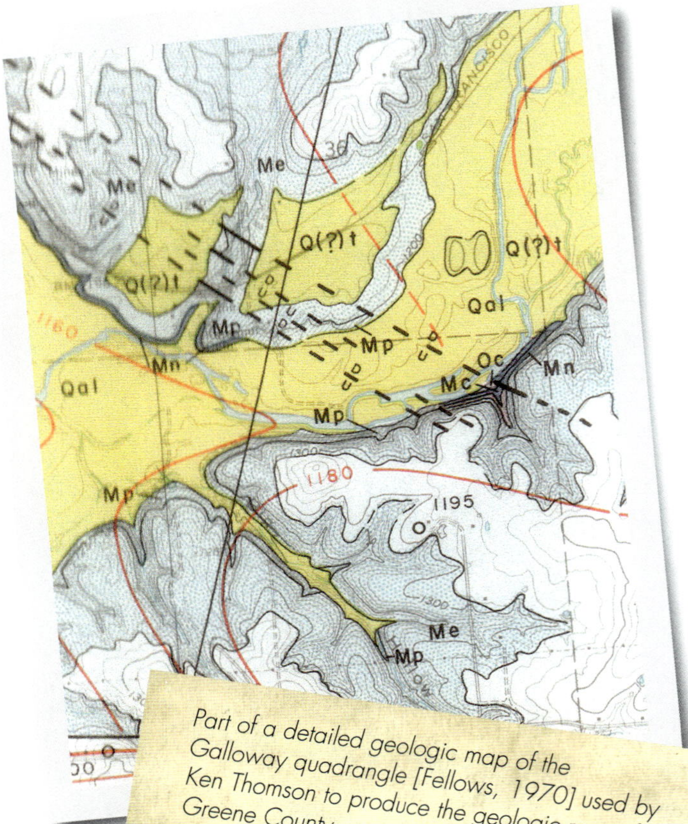

Part of a detailed geologic map of the Galloway quadrangle [Fellows, 1970] used by Ken Thomson to produce the geologic map of Greene County.

stereoscopic aerial photos of the type used in making topographic maps, to help in constructing the geologic map. Experience helps him recognize landforms he knows to be associated with specific rock types, so he can confidently extend his mapping into places where there are fewer outcrops. Subtle clues related to vegetation, bedrock fractures, and drainage patterns are all taken into account. Using all of his skill and insight, the geologist draws in the contact lines separating formations, adds any faulting he has found, and produces a completed geologic map that needs only color to become a work of geological art.

This is essentially the process Dr. Thomson used to produce the geologic map of Greene County, 1986. In doing so, he used previous geologic mapping by numerous authors, credited them for their work, and supplemented that by his own field and office work. Mapping was initially done at 7 ½ -minute scale [2 ½ inches = 1 mile] on topographic maps, which were then reduced to 1 inch = 1 mile scale, and compiled onto one sheet covering Greene County. Thomson's 1986 map has been further reduced as an illustration on page 37-38. The rock unit descriptions beginning on page 18 are also little changed.

All of the maps—outcrop maps as well as completed maps—used in preparing the geologic map of Greene County are on file in the Geologic Map Repository at the State Geological Survey in Rolla.

Outcrop

Outcrop: The term "outcrop" is a favorite of geologists, who love to use it at every opportunity. It is a noun, meaning an exposure of bedrock at the surface, as in a bluff or roadcut. An outcrop does not necessarily mean a prominent feature. It may be very subtle, as a rock exposure in a road ditch, but it must be bedrock in place, not boulders separated by erosion and weathering. The word is often erroneously used as a verb, as "The unit *outcrops* along the bluffline." It should read, "The unit *crops out* along the bluffline." To understand the subtle difference, replace the word "crops" with "jumps." The examples then read, "The unit *outjumps* along the bluffline," whereas the proper usage would read, "The unit *jumps out* along the bluffline."

Outcrops may be few and far between in heavily vegetated areas such as Greene County, where surface soils and residuum obscure bedrock. Where geologic outcrops are plentiful, geologists use solid lines to mark the contacts between different rock formations, and the locations of structural features, but where they are less certain, the lines will be dashed.

Geologic maps are almost always works in progress, subject to revision as newer, more detailed information becomes available. Graduate students are often assigned a quadrangle mapping project as their thesis work for a masters degree. From time to time the names and definitions of rock formations may change, so all geologic maps must be considered in the context of the time they were produced. Thomson's Geologic Map of Greene County, for example, was the compilation of the work of many geologists, plus his own work.

Explanation

Rock Units

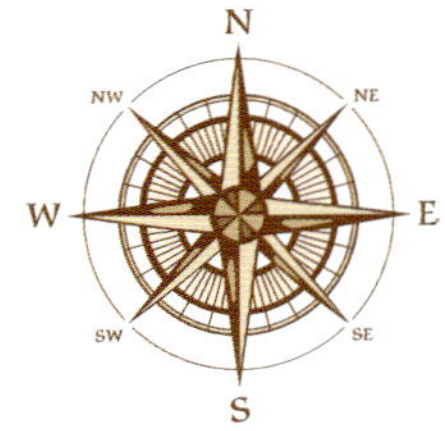

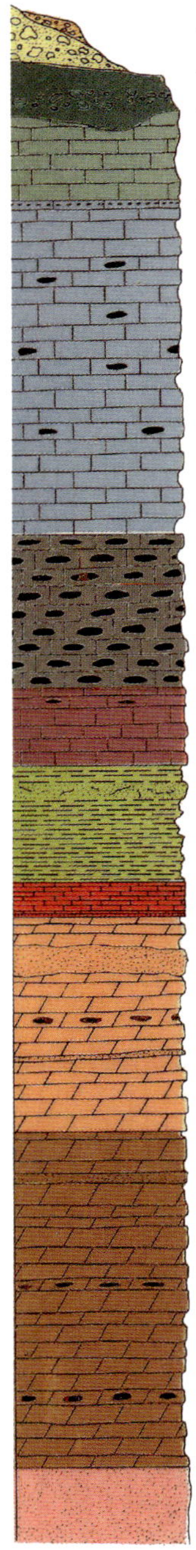

Qt	Terrace Deposits
Qal	Alluvium
PCS	Channel Sands Deposits
Mw	Warsaw Formation
Mbk	Burlington-Keouk Limestone
ME	Elsey Formation
Mp	Pierson Formation
Mnv	Northview Formation
Mc	Compton Formation Bachelor Formation
Oc	Cotter Dolomite
Ojc	Jefferson City Formation
Or	Roubidoux Formation

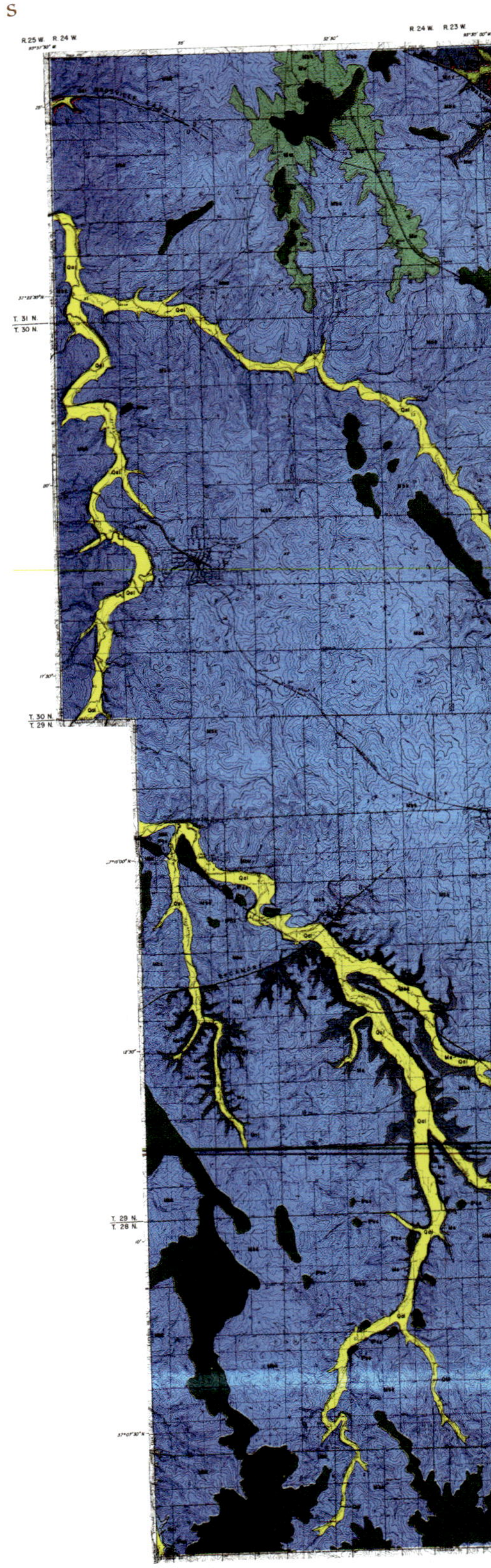

Geologic Map Commissioned by the Watershed Management Coordinating Committee. Field work financed by the Missouri Department of Natural Resources, U.S. Geological Survey (CUSMAP Project), and Southwest Missouri State University Faculty Research Committee. Topographic base map compiled from U.S. Geological Survey topographic quadrangles.

Geologic Map of Greene County

Compiled by Kenneth C. Thomson - 1986

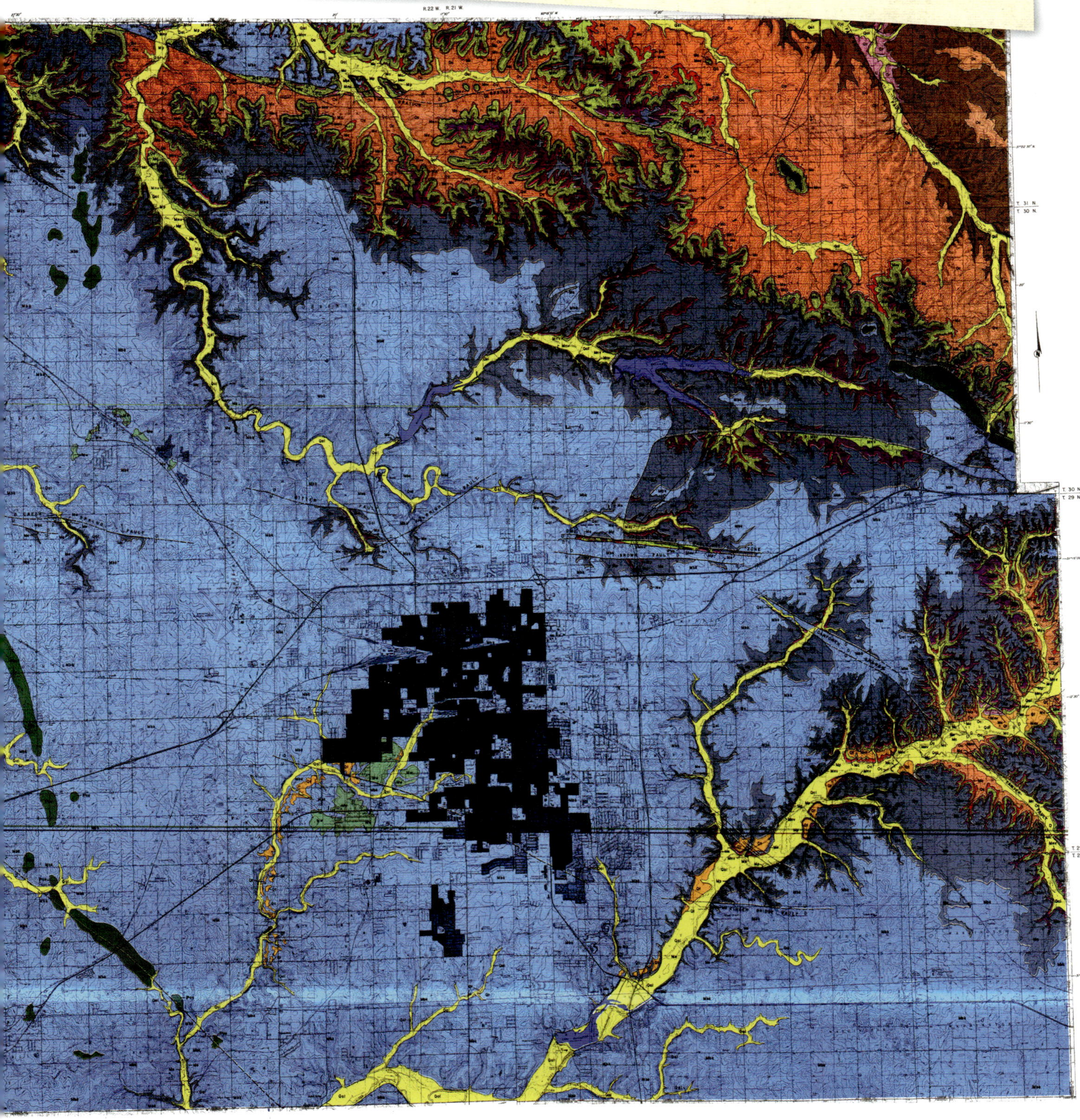

CHAPTER 5

Geologic Structure

The structural geology map of Missouri reveals a strong NW-SE orientation of major fault systems as well as anticlines and synclines.

Geologic structure is what has physically happened to a sequence of rock units since their emplacement. Things like warping, bending, folding, fractures, faults, and other displacements. In areas such as Greene County which is entirely underlain by sedimentary rocks, and has had a relatively uneventful geologic history (as compared to the complexly-folded Ouachita Mountains of Arkansas and Oklahoma), geologic structures are rarely noticeable to the untrained eye. Gentle upfolds (anticlines) and downfolds (synclines) are present, but few people are aware of them. Faults are the most common structural features. A fault is a fracture in rock layers along which there has been movement—up, down, or laterally. Greene County faults tend to be short and occur entirely within county boundaries, but some are regional in extent and cut through several counties. The Bolivar-Mansfield fault trends northwest-southeast and connects the named cities. Parallel with this fault, but on the southwest side of Greene County, is the Chesapeake fault, which can be seen in road cuts on I-44.

Faulting along the south side of a complexly faulted zone at Valley Water Mill Park. In Greene County fault planes are usually obscured by "fault gouge" produced when massive rock sections move against each other.

It is customary to expect earthquakes when the earth shifts along fault zones. The San Andreas fault zone in California is a familiar example. When it moved in 1906, the great San Francisco earthquake and subsequent devastating fire was the result. More recently, the Haitian earthquake of 2010 caused widespread devastation and loss of life. But California and Haiti are both in seis-

Delicate instruments called seismographs record earthquakes that occur when there is movement along geologic faults.

mically active areas. Southwestern Missouri, on the other hand, is in a seismically quiet area. There is no history—not even folklore—of great earthquakes, and while there are occasional tremors, seismicity is not something we lose sleep over. The numerous faults mapped in Greene County seem to be inactive, and no measurable movement has been detected on them.

Earthquakes are felt here, but for the most part southwest Missourians are unaffected by seismic activity. When earthquakes happen, shock waves are transmitted through the earth, and strong quakes often trigger unusual responses far from the epicenter—or ground zero—of the events. For example, the Good Friday Earthquake in Alaska in 1964 caused water levels in Springfield-area water wells to move up and down. Some wells were muddied, and at least one sinkhole collapse (near Lebanon) was attributed to the Alaskan earthquake.

Seismic effects on water wells are well documented; as a seismic shock wave passes through an aquifer it compresses the rock and causes the water level to rise briefly, as the shock wave passes. Water wells equipped with automatic recorders routinely document those events, even though they may originate thousands of miles away. Most recently, the 7.0 magnitude Haitian earthquake was recorded on water well recorders, and some residents actually felt the quake.

The structural geology map of Missouri reveals a strong NW-SE orientation of major fault systems, as well as anticlines and synclines. The ordinary citizen, however, may not notice subtle differences caused by geologic structures. Soil and unconsolidated surficial materials (residuum, alluvium, and glacial drift) overlie bedrock throughout the state, and when vegetation is added, it can be very difficult to spot geologic structures. In fact, many structures can be mapped only indirectly, by using airphoto interpretation and well control. Faults are shown on geologic maps by bold, black lines, implying that one can go to one of the black lines and find a visible break in the rocks. In practice, this is rarely possible. Fault traces tend to be obscured by fault gouge, which is shattered rock broken up when adjacent rock units grind against one another when there is movement along the fault. Careful study, though, allows the field geologist to map faults in considerable detail, including the amount of movement [called displacement, or throw], along a fault. In more than a few cases, faults can be determined by noting different rock strata in contact—as Ordovician dolomite in horizontal contact with Mississippian limestone. Because the two are so different in age, the dolomite would have to be upthrown with respect to the limestone for them to be in horizontal contact.

Faulting is caused by some form of differential force that causes adjacent blocks of the Earth's crust to move. If the force is applied as compression, strata may be bent upward, or one side may override the other. This is called thrust faulting, and is considered relatively rare in Missouri. Tensional forces cause strata to pull apart, a process called normal faulting. Faults in Greene County are thought to be normal faults. There are also faults that involve lateral movement, where one side moves with respect to the other. These are called strike-slip faults.

Greene County also has some more complex structures called horsts and grabens. A horst is formed where there are two parallel faults close together, and the block between them has been pushed up. A graben is much the same, except the intervening space has been dropped down. Finally, a swarm of northwest-trending faults along Pearson Creek in southeastern Springfield was the area where Henry Schoolcraft found "Indians and backwoodsmen" mining and smelting lead in the early 1800s. There were numerous mines and prospect pits in this area. Some of the old tailings piles can still be seen today [NOTE: this is a residential area and permission is required to visit]. A marble plaque near Kershner Spring commemorates Schoolcraft's visit in 1819.

Photolineaments

Geologic structure may also be seen in more detail through stereoscopic analysis of aerial photographs and/or satellite imagery. Careful study reveals numerous straight lines that are presumed to reflect underlying fractures in bedrock. These are called photolineaments (see map on page 43-44). The map shows most of the major faults as well, and the orientation of the lineaments strongly reflects the NW-SE structural grain of the county, as well as a secondary NE-SW orientation. The most intense occurrence of lineaments seems to lie in a north-south band that crosses the county just west of Springfield.

While the construction of a lineament map is relatively easy, it is more challenging to explain why there are so many straight lines, which are considered anomalous in nature. They are thought to be bedrock fractures, though it is not clear how deep they run, nor is it clear whether they form pathways for rainfall runoff to move into the groundwater aquifers beneath Greene County's footprint. Whether lineaments play an active role in sinkhole formation is a logical question, but it is apparent that lineaments occur throughout the county, not just in the sinkhole-prone areas. A new LIDAR map of Springfield and Greene County is in preparation. The new map is expected to reveal photolineaments in much more precise detail, so the existing, 30 year old map may soon be superseded.

Named Geological Structures (Alphabetical order)

1. **Brown Fault**: The structure, just north of Springfield, runs east-west for about 6 miles.

The Burlington Limestone is on the north or downthrown side of the fault, with Elsey and Pierson strata on the south side. Indications of the fault are subtle, including changes in topography, soils and streambeds.

2. **Clear Creek Spring Fault Zone**: Running nearly east-west about 2 miles south of Willard, this 5-mile long disturbed zone has one major fault and two small subsidiary faults to the south. The north side is downthrown. An unnamed spring southwest of Brower Spring is on the fault line. Dipping beds of the Burlington Limestone crop out on the county road west of Brower Spring.

3. **Dadeville Fault**: The fault enters Greene County in the extreme northwestern corner, extending for about 3 miles to the east-southeast. The north side is downthrown. The fault runs west and north into Dade and Cedar counties.

4. **Danforth Graben**: This structure consists of two parallel NW-SE trending faults with the area between them dropped down from 70 to 100 feet. The southeast end of the structure seems to join with the Sawyer fault.

5. **Fair Grove Fault**: Displacing the three oldest rock formations in Greene County–the Roubidoux Sandstone and the Jefferson City and Cotter dolomites–the Fair Grove fault crosses the Pomme de Terre River and extends northward into Dallas County. The east side is upthrown.

6. **Fassnight Fault**: About 7 miles long, this NW-SE oriented fault runs across southern Springfield. It is a "scissors" fault, with the north side upthrown on the south end, and downthrown on the northwest end, indicating a warping movement of up to 50 feet. The presence of the fault is betrayed by elevation differences, well log information, and photolineaments. It appears that the fault may connect with the Kinser Bridge fault to the southeast.

7. **Graydon Springs Fault Zone**: This 35-40 mile long fault zone runs across northern Greene County from just west of Fair Grove Mound, all the way into Polk County. The faulted area is part of the much longer regional structure known as the Bolivar-Mansfield fault, perhaps the most significant geological structure in southwestern Missouri. Displacement along the fault is up to 200 feet, with the north side down. Highway 13 crosses the structure almost 11 miles north of I-44. Driving north on the south side of the fault zone, the road cuts are in Cotter Dolomite, which gives way to downdropped Northview and Pierson, then across the second fault into Burlington Limestone. The road cuts were new in 2009, when the faults were easy to spot. As the area becomes revegetated they will be more difficult to see.

8. **King-Ritter-Fulbright Fault Complex**: A complicated area laced with small faults lies just northwest of Springfield. Hwy. 13 crosses first the Fulbright fault, then the Ritter fault. The King fault is short, about

(continued on page 45)

Fault & Lineament Map of Greene County

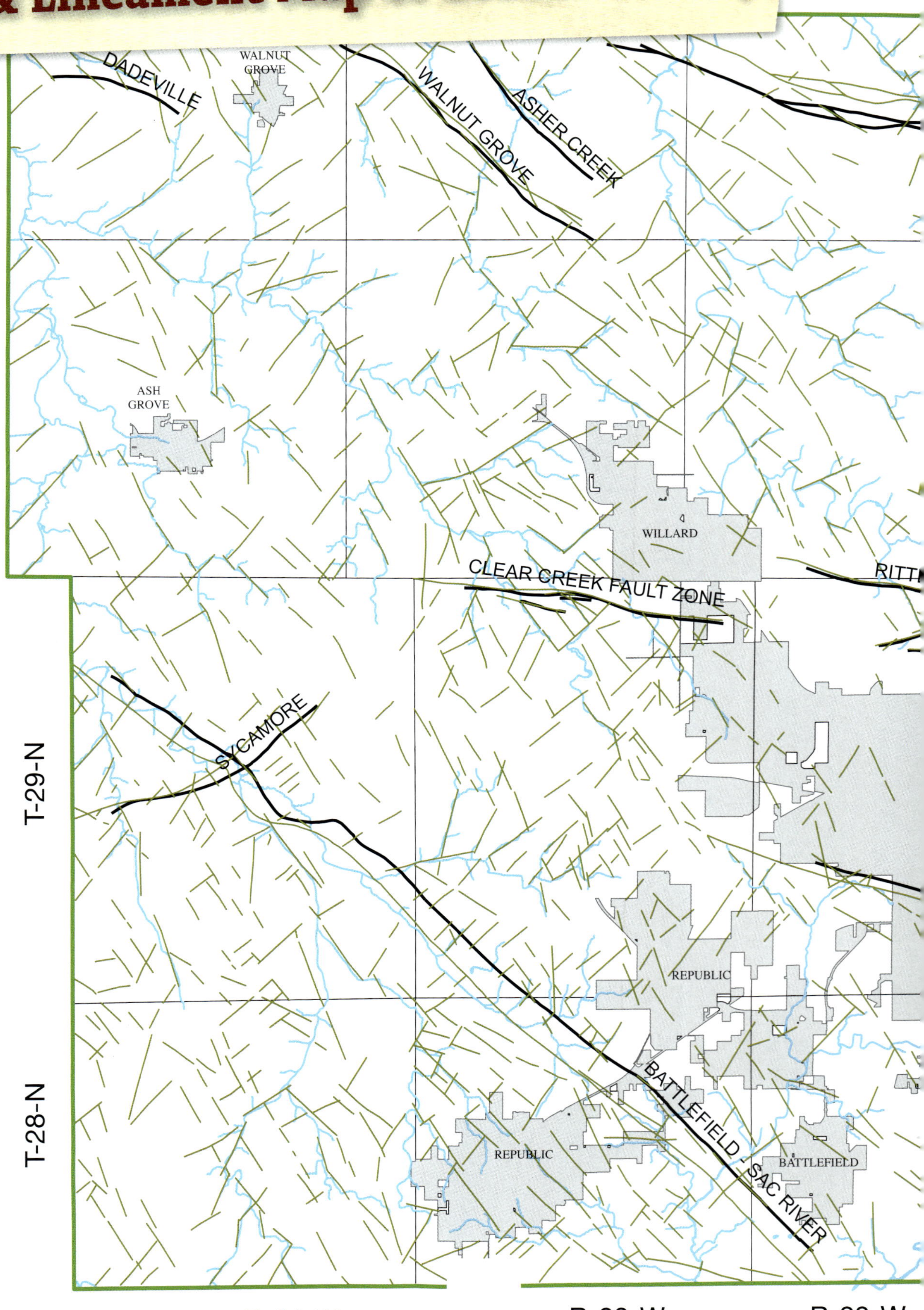

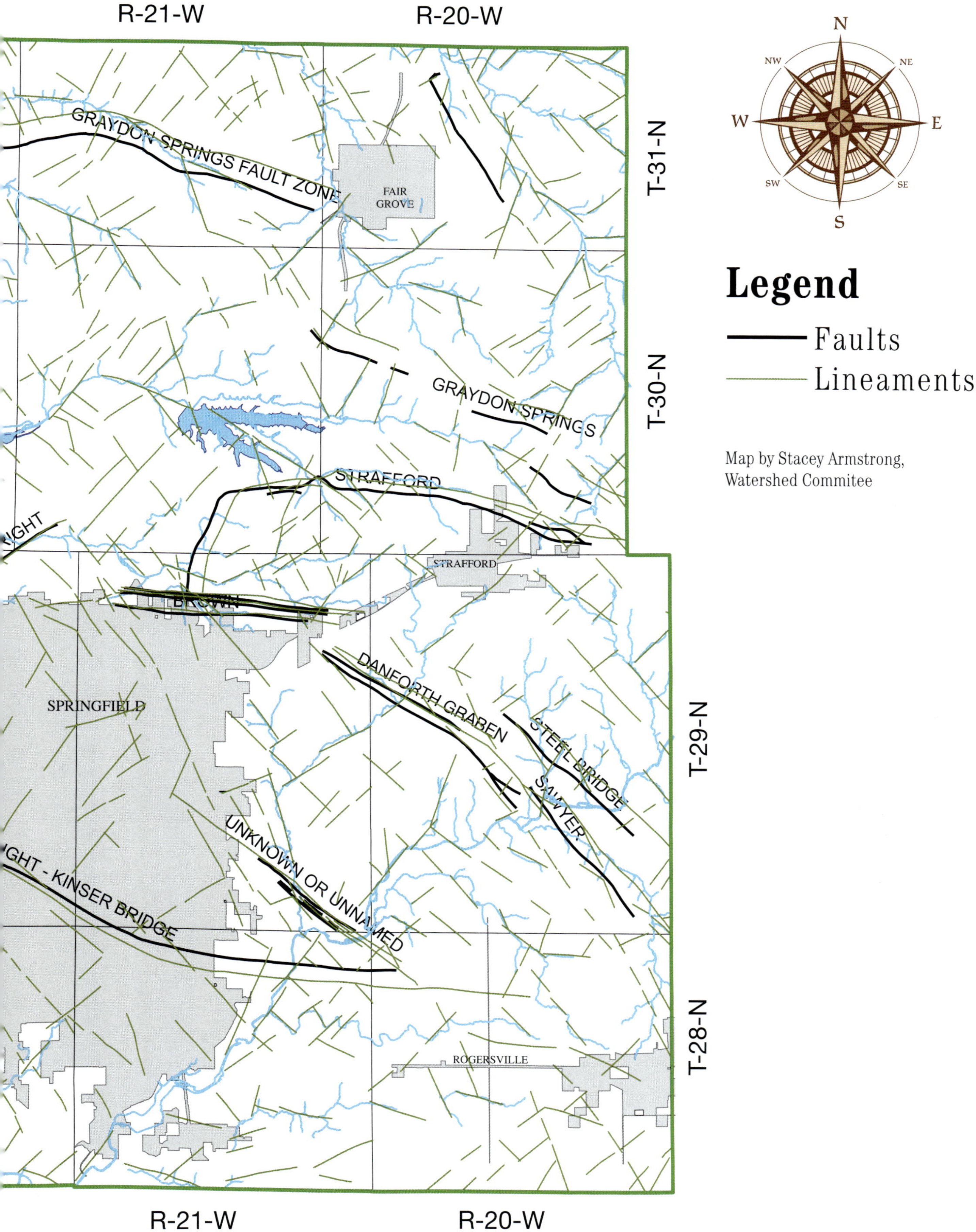
R-21-W
R-20-W
N
NW
NE
W
E
SW
SE
S
Legend
Faults
Lineaments
Map by Stacey Armstrong,
Watershed Commitee
T-31-N
T-30-N
T-29-N
T-28-N
GRAYDON SPRINGS FAULT ZONE
FAIR GROVE
GRAYDON SPRINGS
STRAFFORD
STRAFFORD
BROWN
SPRINGFIELD
DANFORTH GRABEN
STEEL BRIDGE
SAWYER
UNKNOWN OR UNNAMED
GHT - KINSER BRIDGE
ROGERSVILLE
R-21-W
R-20-W

a mile long, and does not appear to cross Hwy. 13, though it does trend eastward. It is upthrown on the north side. The Ritter fault is downthrown on the north side. It is exposed in Smalleys Sinkhole Cave, which is tributary to Fantastic Caverns. The south side of the Ritter fault has been upthrown about 130 feet in the vicinity of Ritter Springs, but the displacement is about 80 feet to the east and west. The Fulbright fault has a displacement of 180 to 210 feet, downthrown on the north side. It is traceable by photolineaments, but difficult to spot in the field.

9. **Kinser Bridge Fault:** Obviously named for its proximity to Kinser Bridge just southeast of Springfield, this 4-mile long fault trends east-west and has a displacement of 12 to 100 feet, north side upthrown. Projecting the fault to the west, it connects with the southeastern end of the Fassnight fault.

10. **Pearson Creek Fault System**: Dr. E. M. Shepard described this faulted area in 1898, but the initial discovery of it dates to Henry Schoolcraft's visit to the area in 1819. Shepard, along with Otto M. Smith and Paul C. Standley, published a report in 1905 in the Bulletin of the Bradley Geological Field Station of Drury College, where Shepard was a geology professor and Smith and Standley were students, class of 1907. They had examined the lead mines along the valley of Pearson Creek, as they were when operating. The Pearson Creek mining area had many "diggings," numerous owners, and was operated intermittently from its pre-1819 beginning to about 1916, when (according to Park, 2005) all mining ceased. A "swarm" of northwest-trending faults provided a favorable setting for ore minerals to be deposited in the rock strata shattered by faulting. Shepard examined the mines in considerable detail, but at the time the stratigraphy of the area was not as refined as it is today. And after 90+ years, the area is obscured by vegetation and modified by residential development so that it is no longer possible to see the area as it was during its mining heyday.

Shepard noted that the ore—mostly galena with some sphalerite—occurred in a "gumbo" consisting of "…white clay," which he said was the product of weathering of the "Hannibal shale." Today we would likely identify the shale as part of the Northview formation. Unfortunately, during the long mining history of the Pearson Creek area, so many mines, pits, and prospect digs were developed that it is difficult to sort them out. Worse, tailings from later diggings were sometimes used to fill in older workings, so once-clear geological relationships are obscured. In modern times, residential development has spread throughout the historic mining area. Nevertheless, one can still find tailings piles, water-filled pits, and old concrete foundations once used for mine buildings and equipment. Projecting the faulted zone to the northwest on the basis of photolineaments suggests the Badger Mine lies in the same fault trend. The tailings pile of the Badger Mine and some of the concrete foundations for the mine workings are still in place, in an unlikely location next to the swimming pool for the Cinnamon on the Hill development. The old mine shaft has been filled.

11. **Sac River-Battlefield Fault:** This long fault enters Greene County's southwestern corner from Christian County, and runs northwest for about 20 miles, into Dade County. The Sac River—especially the Pond Creek branch—seems to follow the fault. The northwest end is crossed by the Sycamore fault, a northeast-trending, 6-mile long structure. Displacement on the fault is about 50 feet, upthrown on the northeast side. Pennsylvanian channel sands obscure about 4 miles of the fault's southeast limb.

12. **Sawyer Fault:** This structure in southeastern Greene County can be traced from the James River southeast along Sawyer Creek valley to the Webster County line. To the northwest it is covered by the James River floodplain, and seems to extend into the Danforth graben. Displacement is 50-60 feet.

13. **Steel Bridge Fault**: First described by Dr. Shepard in 1898, the Steel Bridge fault runs parallel and about 1 ½ miles to the northeast of the Sawyer fault, in southeastern Greene County. Photolineaments are strong evidence for the faulting. Displacement is from 20 to 75 feet, down to the northeast.

Quaking Mound Spring is a small, marshy area in a pasture. In its heyday it was a large, vegetated mound that trembled when one walked on it.

14. **Strafford Fault:** This prominent structure comes into east-central Greene County from Webster County. It trends almost due west for about 9 miles, then turns sharply to the south, and comes to an abrupt end against the Valley Water Mill fault, for a total length of approximately 12 miles. The west and north sides are down. While the Strafford fault is one of the most significant structures in Greene County, it is not easy to trace. Thomas R. Beveridge mapped the fault in the Bassville quadrangle, using subtle clues including the ability to differentiate between cherts from various formations, coupled with photo-lineaments and strata exposed in several mining prospect pits.

15. **Sycamore Creek Fault:** A 4-mile long fault in western Greene County crosses Sycamore Creek about mid-length, then strikes northeast, crossing Sac River and the Battlefield-Sac River fault at right angles, about ¾ mile northwest of the

(continued on page 48)

Fault Finding

Faults are usually shown on geologic maps as heavy black lines (dashed where inferred). In a perfect world, one might expect to go to the marked place in the field, and stand astride the fault with one foot on the upthrown side and the other on the downthrown side. But in Missouri, where vegetation, soil and residuum almost always cover the bedrock, it is rarely possible to actually touch the interface where the movement occurred. Instead, geologists look for indirect evidence of faulting—shattered rock or "fault gouge," differences in rock types on opposite sides of stream valleys, and differences in elevation of a specific marker bed—all good evidences that faulting has occurred.

When there is vertical or horizontal movement along a fault, the usual result is an earthquake. The earthquake shocks are generated when the rock masses on both sides of the fault try to move against each other. Typically there is a lot of rock breakage and deformation along the fault plane. More often than not, this creates a disturbed zone that conceals the actual fault trace. The disturbed zone is likely to be most pronounced at the surface. Sometimes faults are encountered in mine workings or in caves, where they tend to be more evident.

In this book, the descriptions of faulting are summarized from Dr. Thomson's original report. The faults documented in Greene County are generally thought to be inactive. That is, movement ceased long ago and the

Looking west across the north end of Valley Water Mill Lake, where several parallel faults define a structure called a "horst," where strata between the faults has been thrust upward. The entrance to Wasp Cave is on the center right of the picture.

county is now considered to be seismically quiet, a low-risk area for earthquakes. Unfortunately, there is no guarantee that movement along any mapped fault may not resume.

Find it hard to imagine how a sequence of hard rocks some 2,000 feet thick can be bent, folded, broken, and smashed? Think globally! The Earth is a planet about 8,000 miles in diameter. It is dynamic, ever-changing, with volcanism, plate tectonics, landslides, floods, meteorite impacts, and various other dislocations constantly happening. Springfield's "footprint"—2,000 feet of sedimentary rocks—is miniscule when one considers that it is only 9 ten-thousandths of the distance from the surface to the center of the earth. This would be comparable to the thickness of a coat of paint on the roof of the Hammons Tower in Springfield! The wonder is—given the span of geologic time—there aren't more structural effects.

Stonebridge karst complex. Displacement is from 50 to 170 feet, down on the northwest side. The faulted area near the confluence of Pickerel Creek and the Sac River has an interesting mining history. John Park's Missouri Mining Heritage Guide (2005) notes that in 1887 John McDaniel discovered galena crystals in Prospect Spring, just across the creek (west) from the old Prospect Church Cemetery [the church building was burned by vandals]. Thereafter, a mine was opened and about 200 exploratory shafts were sunk in the area, the deepest of which was said to be 62 feet. Production was reportedly "about 200 railroad cars of ore...mostly lead with some zinc." On the floodplain just north of Prospect Spring is the Quaking Mound Spring, where Shepard (1898) said a large mound of vegetation "quaked" when one walked on it. The site today looks greatly different; the mound of vegetation is gone, and the site resembles a small pond or bog. Clearing the land for agricultural use likely caused the mound to dry out over time.

16. **Valley Water Mill Horst:** Arguably the most interesting and most accessible geologic structure in Greene County is a complexly-disturbed zone in Valley Water Mill Park, just west of the Watershed Center, now under construction. Two parallel faults trending almost east-west cross the upper end of Valley Water Mill Lake at the west end of the faulted area. Here, the Northview formation has been pushed up between the two parallel faults, into contact with the Burlington Limestone, forming a horst. [The opposite effect, where the central space is downdropped, is called a graben] Looking across Valley Water Mill Lake, the structure appears to be a graben, because the central part is lower than the sides, but the Northview is much more easily eroded, so the more resistant Burlington appears dominant. The perimeter walking trail around the lake has a short spur leading to an overlook where the faulting is clearly exposed. At this overlook one can actually touch the fault plane. The intervening space between two faults is concealed by vegetation, but when that is cleared away, the Northview is exposed. Following the horst eastward, it crosses Hwy. 65 at Valley Water Mill Road, then continues eastward for about 2 ½ miles before fading away.

CHAPTER 6

Mineral Resources

Non-metallic mineral resources–often called industrial minerals because of their association with construction and manufacturing–are a major contributor to the region's economy.

Greene County has a rich mining history dating from the early 1800s and ranging from lead, zinc, and iron metals through coal, clay and shale, sand and gravel, and limestone. Metal mining began before 1819 by native Americans and early settlers, as noted in the writings of Henry Schoolcraft, for which the Schoolcraft Expressway (Hwy. 65) is named. Several mining camps or "diggings" were worked in Greene County, but the activity was sporadic and came to an end in the early 1900s. Non-metallic or industrial mineral production, however, has continued to increase and limestone is currently the only mineral commodity produced. Fortunately, limestone is plentiful in the county and existing quarries have reserves adequate for many years of extraction. Concurrently, environmental regulation related to industrial mineral production has kept pace with growing consumption, and limestone production is likely to continue far into the future,

Train coming out of Springfield Underground on a snowy day. Note shotcrete sealing of Burlington Limestone at the mine portals. Photo courtesy Springfield Underground.

as southwest Missouri grows in population as a regional center of commerce.

There is almost no limit to the amount of limestone available in the county, but the likelihood of additional quarries being opened is probably very low, given the constraints related to the cost of land, zoning regulations, and obtaining operating permits. Public opposition to the dust, noise, effects of blasting, and heavy truck traffic associated with mining and quarrying, is a given, so it is fortunate that the four limestone quarries/mines now operating in Greene County have sufficient reserves for the foreseeable future. There is strong local demand for architectural or landscaping stone, that relies in large part on the weathering characteristics of limestone and dolomite. The Burlington Limestone has a very uneven weathering surface, as spectacularly displayed in the gargoyle-like rock outcrops lining major thoroughfares in the city and county. Any new road project or building excavation is likely to produce piles of weathered boulders, which are much prized by property owners who want "a piece of the rock" to beautify yards, grounds, and parking lots. Some of the city's subdivisions have elaborate rock gardens and water features that complement the architecture and add pleasing components to the cityscape.

While the supply of landscaping stones from the Burlington Limestone is plentiful, rocks are also imported from other areas. The Jefferson City-Cotter Dolomite yields weathered stones with a different "look" that are also highly popular. An example of their use is found in the flower beds that grace the main campus of St. John's Hospital. In earlier years, smaller fieldstones were used for buildings and fences in public areas such as Fassnight Park, the spiked stone fence around the old Bluff School, and in numerous other locations.

Landscaping stones of Jefferson City Dolomite grace flower beds on St. Johns Hospital main campus, Springfield.

Mineral resources are commonly discussed as either metallic (gold, silver, iron, lead, zinc, etc.), or non-metallic (limestone, industrial sand, clay, gypsum, etc.). Greene County has a long history of both metallic and non-metallic mineral production, but no metallic ores have been mined in the county since the 1930s. Non-metallic mineral resources—often called industrial minerals because of their association with construction and manufacturing—are a major contributor to the region's economy. Lead and zinc ores were mined in the early 1800s and into the 1930s, but production levels never approached those of the Tri-State District around Joplin. Iron minerals are present at 18 localities in Greene County, but none of

Galena, a primary ore mineral of lead. Galena is lead sulfide (PbS), and when it is smelted, the sulfur is released in stack gases, which combine with precipitation to form sulfuric acid, widely blamed for "acid rain." Scale is in inches.

the deposits are large enough for sustained commercial production.

Lead and Zinc

Few people in modern Springfield/Greene County know that the area was a focus for lead/zinc mining more than a century ago. Mining camps sprang up near Brookline, Ash Grove, Pickerel Creek, and along Pearson Creek. At the time, there were no environmental regulations, so it was possible to mine the ore and smelt it onsite, before the harmful effects of lead became known. Lead comes from the mineral galena, known by its chemical name, lead sulfide (PbS). Galena is a silvery mineral that looks like lead metal, but is very brittle. In order to make the familiar malleable lead metal used in fishing sinkers, car batteries, plumbing, and other common applications, the sulfur must be driven off by smelting, or heating the ore to a high temperature. This process drives off the sulfur as a gas, leaving lead metal.

Today, strict environmental regulations apply to lead smelters. No lead smelters remain in southwest Missouri, but there are two operating smelters in eastern Missouri, serving the mines of the Viburnum Trend, once the most productive lead mining area in the world. Extensive cleanups are underway in the Joplin area and in Missouri's "Old Lead Belt" around Bonne Terre-Park Hills, to remediate old mine tailings and other instances of lead contamination. In Springfield, some limited contamination associated with the old Badger Mine has been remediated. There are still some waste piles associated with mining in the Pearson Creek area of southeastern Springfield. Some of these may require attention in the future, as urban growth continues.

The Ash Grove "diggings" began in 1859 with the discovery of ore-grade mineralization in a water well. The area became known as the Corum Diggings, but ore production was minor. Ore was discovered on the Duncan property in 1888, exploited by several companies and several mines were opened. The Golden Eagle mine is said to have produced 400,000 pounds of lead. Whether this was lead ore or lead metal is not clear, but other mines nearby were opened by the Gulf Company and the Clinton Company. About 1889 the Taylor and Edgington ore body, lying at a depth of about 75 feet, led to the opening of several mines, having tunnels with the ore body extending for some 600 feet. Several shallower mines at the nearby Getty Diggings also produced lead and zinc.

The geologic map shows no structures passing through the various diggings, so the discovery of ore minerals must have been through astute observations at the surface or in excavations for wells or buildings. The geologic map shows the area to be entirely underlain by the Burlington-Keokuk Limestone. Absent the discovery and proving of a multi-billion dollar ore body in the Ash Grove area, it is highly unlikely that any lead/zinc mining will occur in this area in the future. However, it would not be out of the ordinary—in the course of road or building construction—to find small deposits of lead and zinc minerals such as galena and sphalerite, perhaps associated with other minerals of similar chemistry. These would have value as mineral specimens, but not as ore minerals for metal production.

Native Americans were mining lead along Pearson Creek in 1819, when Henry Rowe Schoolcraft visited the area and wrote about his findings. Mining and processing of lead ores occurred sporadically in the Pearson Creek area for almost a century, finally ending about 1916. Among the numerous mine owners/operators were two former governors of Missouri, Gov. Joseph W. McClurg (1869-1871), and Gov. John S. Phelps (1877-1881). Park (2005) details the complex mining history of the Pearson Creek diggings. An archaeological and historical survey of the mining area was done for the City of Springfield (Cooley, et al., 1976). It details the locations of numerous mines and mine workings, along with Schoolcraft sites, springs, and historic cemeteries.

The mining area is along a series of closely-spaced, northwest-trending faults. Mineralization was primarily in the Northview formation, in fissures or crevices related to the faulting. The lead and zinc ore reportedly occurred in "gumbo" or "tallow clay," which was difficult to separate from the ore minerals. Over the long history of mining, numerous shafts were sunk, to depths of up to 140 feet. Drifts as long as 1200 feet were opened and mined at various times. Some of the ore was smelted on-site, using primitive smelters constructed of sandstone slabs and fueled by hardwood logs. Production statistics are imprecise at best, usually consisting of estimates of the value of unprocessed ore, or noted as "xx railroad carloads of lead ore," with no indication of the price paid for the ore.

Today there is little evidence remaining of the long, colorful history of mining on Pearson Creek. There are a few piles of tailings from various mines, some shallow prospect pits, and old concrete foundations for mining equipment and buildings. The Copper Mill subdivision forms the eastern boundary of what can still be seen of the mining area. Ironically, no copper was ever recovered from the mines. The

area where most of the old mines are now is mostly residential, and several large estates are gated. Privacy signs are prevalent, but the area has been designated as a Greene County Historic District, so respectful requests for access permission are recommended; do not trespass.

The Pickerel Creek Mines were in an unlikely place near the old village of Plano, in western Greene County. John McDaniel reportedly found galena crystals in a spring in 1887, and subsequently a mine opened and production of lead ore began. The mine operated for several years, producing "about 200 rail cars of ore." There were extensive diggings in the area; one shaft reached a depth of 62 feet, and there were said to have been some 200 exploratory shafts. Thomson (1986) described the district as follows: "The shafts are located along three parallel fissures which strike approximately N. 25 W. These seem to extend for about 1/3 mile and are fairly close together. The western fissure is entirely within the NW ¼ of section 34 where a crevice 6 feet wide was found in the limestone. A shaft was sunk to a depth of 40 feet and a 62 foot shaft was sunk near a spring in the NW corner of the property. A third shaft was sunk for 46 feet on the top of the hill." The spring mentioned by Thomson does not appear in Bullard's Springs of Greene County.

The Brookline Mining Camp featured several small mining ventures beginning about 1873 and continuing to about 1887. These "diggings" were in an area southeast of Brookline, laid out in a north-south orientation, with the Potter diggings to the north, the Armstrong diggings next, and the Old Silicate Diggings ¼ mile farther south. The Potter property seems to have been the most active. The mine operator, Bay Wilson, produced ore from several shafts and smelted it in his own smelter, on-site. Wilson reportedly produced 300,000 pounds of ore, but there are no records showing how much lead metal the ore yielded. Mining on the Potter tract continued off and on for a number of years, producing "a large amount of ore."

Limonite, a low-grade ore mineral of iron. Minor occurrences of limonite in Greene County are too small to offer potential for future iron mining.

Mines on the Armstrong property apparently were filled sink deposits. Historical accounts indicate that the Brookline diggings were unsafe, in "soft ground," and subject to frequent caving. This suggests that the mines may have been in old sinkholes, or simply in deep residuum; the deepest shafts were 85-95 feet deep, and most of the mining reportedly was at depths of 25-50 feet. In any case, the old mine workings are not apparent today. However, ore occurrences in sinkholes suggest that other shallow deposits of lead and/or zinc ore minerals may be found in future excavations as development in the Brookline vicinity continues. In fact, records maintained by the Missouri Geological Survey document numerous minor lead and zinc mineral occurrences in Greene County.

Miscellaneous Mineral Commodities

Iron Oxide

The minerals limonite and hematite are frequently called "iron ores," but in geological parlance, an "ore" is a mineral that occurs in quantities and sufficient quality that it can be refined into metals. Small specimens of limonite and hematite are commonly found in Greene County, but never in sufficient quantities to be actually mined. Iron minerals are mostly associated with the Pennsylvanian Channel Sands of western Greene County. These are classed as residual deposits and do not rise to the level of ores.

Coal

Small and widely dispersed coal deposits have been found in western Greene County, usually in or associated with sinkholes. Some small-scale coal mining occurred near Brookline in the distant past.

Clay and Shale

Greene County has considerable reserves of clay and shale that have been used in the past to make bricks. However, there is no current production and Thomson (1986) could not find any evidence of clay pits west of Willard that were mined for brick making. Statistics on brick production are available up to 1898; no production has been recorded since then.

Shales from the Northview Formation constitute a significant potential in northeast Greene County for future brick production.

(continued on page 57)

Mineral Resources Map of Greene County

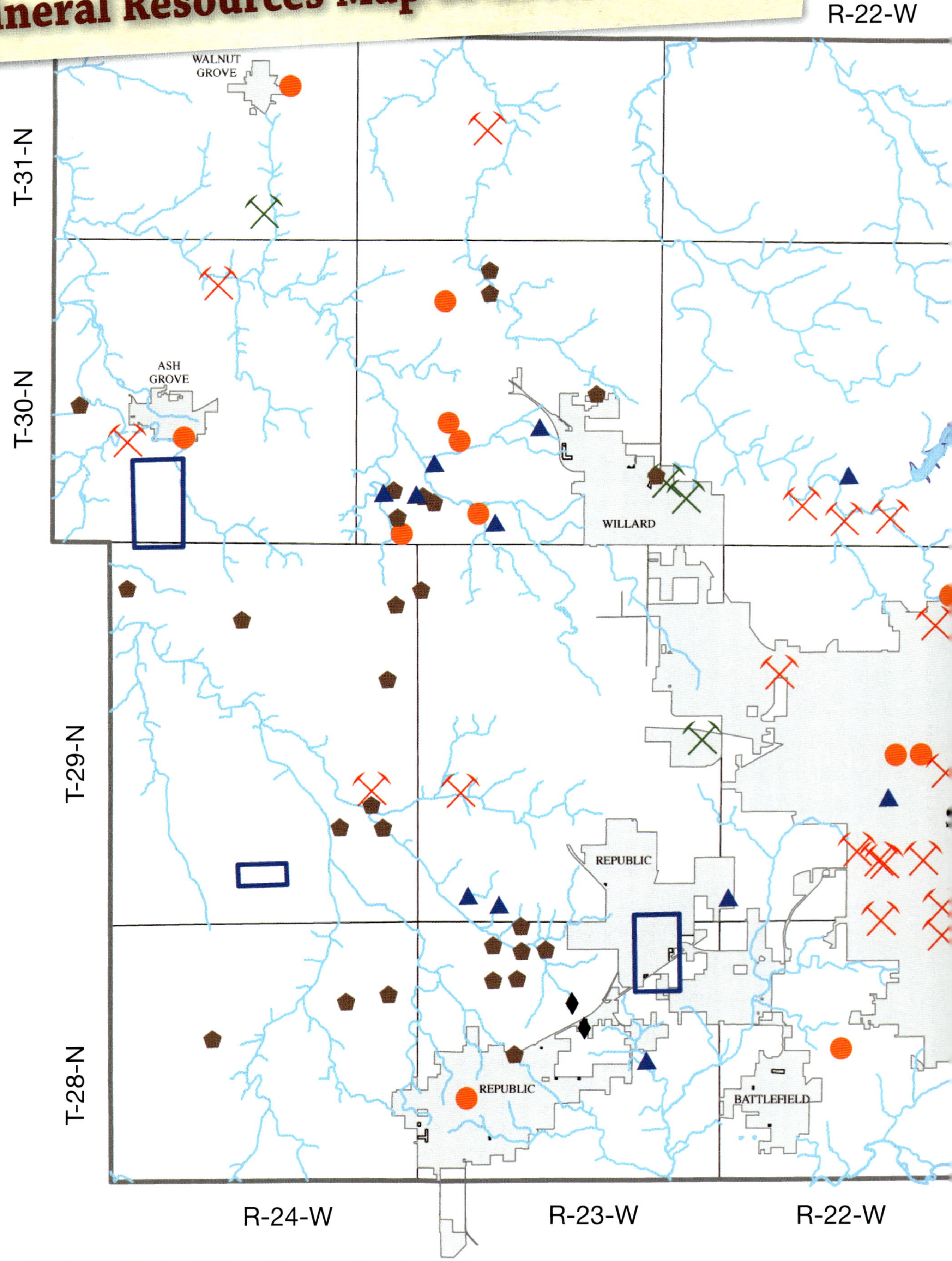

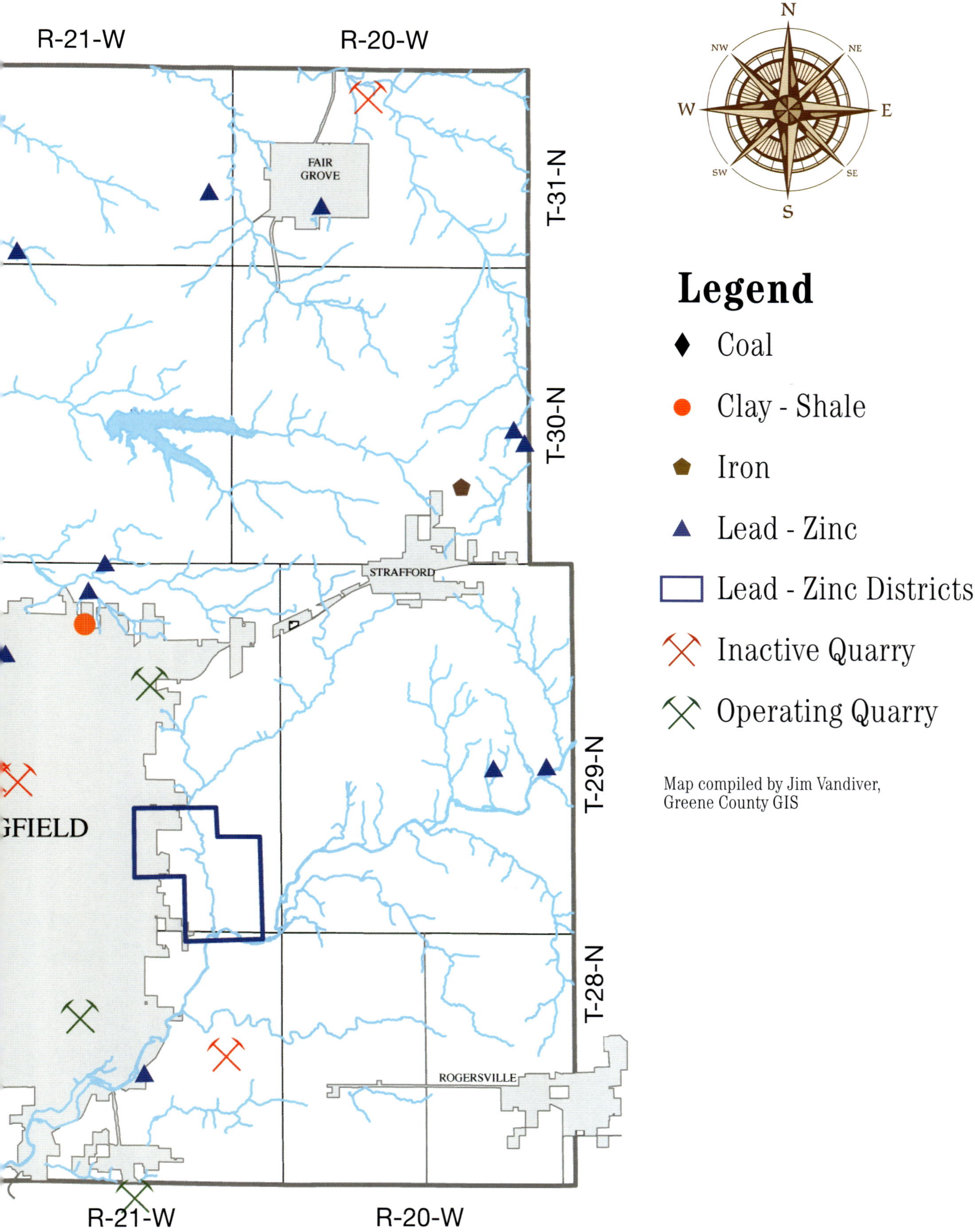
R-21-W
R-20-W
FAIR GROVE
STRAFFORD
GFIELD
ROGERSVILLE
T-31-N
T-30-N
T-29-N
T-28-N
R-21-W
R-20-W
N
NW
NE
W
E
SW
SE
S
Legend
Coal
Clay - Shale
Iron
Lead - Zinc
Lead - Zinc Districts
Inactive Quarry
Operating Quarry
Map compiled by Jim Vandiver,
Greene County GIS

Shale in the Northview Formation is a potential economic resource in parts of Greene County where it crops out. Unlike limestone or sandstone, shale is easily penetrated by a geopick.

The formation is 40 to 80 feet thick. However, well-established brick-makers in neighboring states are currently filling market demands, so there are no current brick producers in Greene County. Historic production up to 1898 was 33,700,000 bricks.

Limestone

Greene County is especially fortunate to have abundant, readily available resources of high-quality limestone, with efficient and responsible production facilities. Limestone is the primary raw material for making cement, while crushed and broken limestone is the key resource for concrete aggregate, base stone for roads, parking lots, driveways, and a host of other applications. Crushed stone mixed with asphalt is essential to the region's road network, while the ubiquitous "white rock" is a familiar sight in nearly every construction project.

The Burlington-Keokuk Limestone is the only formation quarried in Greene County. The stone averages 99% calcium carbonate, with very minor amounts of aluminum, iron, silica, and magnesium. Limestone is heavy, tipping the scales at 165 pounds per cubic foot. A cubic yard of limestone weighs more than two tons. The primary non-limestone component of the Burlington is chert, a rock composed of silicon dioxide that occurs in thin lenses and nodules in the limestone. Fortunately, the best limestone occurs in the upper part of the Burlington formation, where there is less chert. Quarry operators use various strategies to remove the chert in the production process, in order to meet strict quality specifications.

Limestone mining and quarrying [a quarry is an open-pit, surface operation, whereas mining is done underground] are heavy industries that inherently are dusty and noisy. The breakup of limestone to produce such basic

The massive primary crusher for Conco's Willard Quarry is built against the highwall of the sprawling quarry on the south edge of Willard. The quarried rock is the Burlington-Keokuk Limestone.

Highwall in the Willard Quarry exposes typical Burlington-Keokuk Limestone; note small white chert exposures, and the wall section eroded in relief by groundwater solution

commodities as aglime and concrete aggregate requires drilling, blasting, and crushing. Moreover, both the raw materials and the finished products usually move by heavy trucks. Limestone producers customarily have associated ready-mix concrete facilities that serve as a direct link between source material and end product. However, industrial complexes and residential/commercial areas make uneasy neighbors, so modern limestone producers comply with regulations designed to minimize the negative aspects of production and distribution. This is well demonstrated in Greene County, home to four major limestone operations and a number of fabricating and delivery companies. The impact of cooperative work between industry and regulatory agencies has been toward minimizing environmental pollution, delivering cleaner air, and ensuring worker safety.

Thomson (1986) documented 27 quarries in Greene County, all but four of which are no longer in operation. Recently, Springfield Underground opened a new quarry near the Springfield-Branson International Airport that is the first new limestone-producing facility to be opened in the last quarter century. Surface workings are now under development that will lead to an underground mine much like the existing facility on the east side of Springfield.

Going underground is more costly than surface mining, but it minimizes some of the most undesirable effects of stone extraction. First, the noise and dust associated with the mining process are confined underground. Second, the effects of blasting are less apt to generate complaints when they are kept to a working face in the mine. Even so, modern blasting technology that utilizes electronic equipment to break the rock with a succession of millisecond delays is a far cry from the days when a box with a T-shaped handle was the customary way to set off a huge explosion. Limestone quarrying began in Greene Co. at least as early as 1880, when a quarry was opened at Ash Grove. The business was successful, and soon became a new company named the Ash Grove White Lime Association, formed to manufacture lime for use in masonry mortar and for making plaster, used in interior walls before the advent of "sheet rock" (also called gypsum board). To convert limestone into lime, crushed limestone is heated in a kiln to drive off carbon dioxide, leaving calcium oxide (CaO). The Ash Grove Company had two lime kilns in 1880, and had expanded to 11 kilns in 1931, when the quarry closed, having exhausted its stone reserves.

The Ash Grove Company had developed other quarries, and became a large corporation, operating plants in several locations, including the Galloway quarry in southeastern Springfield. In 1904, lime produced at Galloway won top honors at the St. Louis World's Fair, vetting the company's slogan, "Best on Earth."

Quarrying Burlington-Keokuk Limestone, Conco's Galloway Quarry, Lone Pine at Galloway Village.

The Ash Grove Company's Galloway lime plant was a Springfield landmark on South Lone Pine, across from Sequiota Park, until 2008, when it was demolished. Its production facilities had become inefficient compared to other plants the company owned, so rather than upgrade the old plant, it was taken down and the site will be used as green space while the limestone quarrying operation continues. Crushed stone is marketed locally through Conco's ready-mix plant and for other customers.

Ash Grove is also the parent company for Conco's sprawling limestone quarry at Willard. This operation utilizes a thousand-acre site. The modern, highly-mechanized production facility uses computers and GPS systems to organize production of stone products including aglime, various sizes of crushed stone, aggregate for concrete and asphalt paving, manufactured sand, and stackable blocks for landscape architectural applications. Stone is blasted from quarry walls using a bench-based system, hauled to a primary crusher in the pit, then moved by conveyor belts to the surface facility. Here, a complex of humming secondary crushers, screens, and conveyor systems—all controlled by computers from a central control tower—moves the crushed stone—sorted by size—into a series of stockpiles.

The quarry also produces "manufactured sand," used to replace river sand in many concrete mixes. River sand must be imported from river sources. For example, sand from the Missouri River is recovered by sand dredges, then shipped to Springfield.

Manufactured sand is far cheaper because it is produced here and has low shipping costs. The difference is that individual grains of manufactured sand are sharp-edged and angular, whereas river sand grains are rounded by virtue of their transport by the river. River sand gives a much smoother finished surface—such as in kitchen countertops—that manufactured sand cannot match. Cost of shipping is the critical factor; the per-ton cost of shipping river sand from Kansas City to Springfield exceeds the cost of the sand itself. Recycling is a growing factor in today's limestone industry. Asphalt mixes used on roads and parking lots may be up to 80 percent crushed limestone, so a growing trend is to crush old asphalt paving and use it as aggregate for new paving projects. The mechanical planers used to smooth uneven roadways generate waste that can be screened and re-used. Another growing trend is the use of portable crushing/screening units on major construction projects that involve blasting and removal of bedrock. Often these materials can be used elsewhere in the project where crushed stone is needed.

The Willard quarry [formerly Greystone] operated by Conco has been in operation for many years. In the past, resource agencies such as the U.S. Bureau of Mines kept detailed statistics on mineral production, but the Bureau was abolished about 25 years ago and such statistics are difficult to come by. For the years 1952 through 1977, total limestone production in Greene County was 27,916,000 tons, valued at $52,502,000, or a little less than $2/ton. Today, with basic crushed stone selling for about $10 per ton, Greene County's annual stone production is worth several tens of millions of dollars. The numbers would be much higher if the stone came from, say, Joplin or Lebanon. Because most stone is moved by dump truck, when a truck moves 20 miles the cost of shipping equals the cost of the material, so moving it 40 miles would double the cost to customers.

Heavily eroded surface of the Burlington, stripped of overburden and ready for quarrying, Conco's Galloway Quarry.

Griesemer Stone Co. opened a limestone quarry in 1946, at a site east of Springfield that is now Hwy. 65 and Kearney Street. Open-pit quarrying of the Burlington Limestone continued until 1954, when the company went underground with a bold plan to use the space created by mining to build underground warehouses. They adopted a mining plan using the room-and-pillar mining process, with 30 x 30-foot stone pillars on 80-foot centers, leaving a span of 50 feet between pillars. By building concrete block walls between pillars, it was simple to create

Underground space created by mining limestone can be easily developed for use as offices, warehousing, and other uses. Underground "buildings" are easily made by erecting concrete block walls between pillars of solid rock left after mining. Note utility lines attached to the mine overhead. Photo courtesy Springfield Underground.

The visionary plan worked well enough that by 1960 the company had 250,000 square feet of space to lease. In 1984, Griesemer Stone Co. and Security Terminal Corp. merged into a new company called Springfield Underground. The new company continued to mine and market limestone while creating versatile space for other purposes. In effect, the plan eliminated one of the most intractable issues related to mining and quarrying—what to do with the big holes left after mining is completed? Moreover, mining is a one-time operation, but once the mined space has been developed, it can provide an income stream far into the future.

Springfield Underground now has 2.4 million square feet of mined space available. Currently, all the space is leased to various companies for a variety of uses. Rail services go directly

underground "buildings" for warehousing and other uses. The thick Burlington Limestone at the site was very high quality and generally free of fractures, mud seams, and low in chert. It was possible to create rooms with ceiling heights of 30 to 35 feet. By paving over the uneven floor left by the mining process, bringing in water, sewer, and electricity, and adding sprinkler systems for fire safety, the underground space cost roughly two thirds the cost of surface buildings. And because the average temperature of the mine is about 58 degrees, there is little need for heating and cooling for worker comfort. Best of all, the underground "buildings" were completely isolated from tornadoes and floods, and security was easily maintained because the mine portals—the only entrances—could be easily monitored.

Runners compete in a 5K charity run in Springfield Underground. Note long, high-ceilinged corridors in Burlington Limestone. Concrete block walls between rock pillars make it simple to construct underground "buildings" for warehousing, offices, and other uses. Photo courtesy Springfield Underground.

underground, so train cars can be loaded or unloaded at the doors of underground buildings. Trailer trucks and all service vehicles also have access through the wide, high aisles of the underground complex.

The old Galloway Lime Plant, south Lone Pine across from Sequiota Park, was a landmark for decades until it was demolished by its owners in 2008, ending a long history of lime production in Springfield.

Maintenance of mined space is comparatively inexpensive because the underground buildings are not exposed to weather. Energy requirements are also low. Above-ground temperatures range from well below zero to more than 100 degrees F., but the underground temperature is stable at about 58 degrees F. Because of the insulating effects of large rock masses, space can be heated or cooled, and once the desired temperature is reached, it can be maintained economically.

Mining in parts of Springfield Underground continues to generate more space, isolated by distance and temporary curtains from the developed parts of the mine. The company has annual production of around 800,000 tons, with ample reserves for many years into the future.

While mining and space utilization continue underground, the land over the workings can be used for other purposes. During the past several years, land overlying Springfield Underground has been used for the unique "I Love America" Fourth of July event sponsored by the James River Assembly. This patriotic tradition attracts tens of thousands of people, and features tours of Springfield Underground, offering area residents a rare opportunity to experience what it is like to work underground. They also get to see an amazing underground industrial facility, humming with activity, serviced by both trains and trucks, unaffected by surface weather including tornadoes, snowstorms, floods, and heat waves.

Conco's Galloway quarry operates from surface workings, but some past production was underground. The company has two levels of mine tunnels under the northwestern part of its property, near Springfield's busy Glenstone Avenue. The space is currently undeveloped and used to store bottom ash from the old lime plant, but in the future it could be cleared out and developed as underground space.

The landmark Galloway Lime Plant had been prominent along south Lone Pine for decades, blocking the view of the quarry from the

Working face in Burlington Limestone, cut by wire-line saw. The cuts left a wavy surface using special sand for grit to cut through the relatively soft limestone.

street. When the old structures were imploded in 2008, the large scale of the limestone workings became apparent. Today the quarry still produces crushed stone, using in part a business plan designed to more completely recover stone left behind in earlier mining plans, and to clear away kiln residues—which had been stored on site—from the old lime plant.

Phenix Quarry

Steeped in history, the Phenix Quarry in northwestern Greene County was perhaps the most prestigious stone-producing site in Missouri during its heyday. The resource was a marvelous deposit of high-quality, almost chert-free Burlington Limestone. It was quarried, sized, polished, and marketed under the name "Napoleon Grey Marble." [In geological parlance, "marble" is a metamorphic rock, but for marketing purposes, the term is sometimes applied to limestone that is capable of taking a high polish.] Phenix stone was used in many important buildings of the time, including Missouri's state capitol, the Greene County Courthouse and other local buildings, as well as important structures in Chicago, New York, and on the west coast. Fifty rail-car loads of marble went into the Russ Building in San Francisco, built in 1929. Phenix stone also was used in the New York Stock Exchange building and the Petroleum Science building in Los Angeles. The Phenix Company built a town for its workers, complete with an impressive music hall, homes for the workers, and even hired a "professional" minister to service the spiritual needs of its labor force.

The company began as a producer of lime, which was the basis of plaster, the predecessor of "sheet rock" that is in common use today for interior walls of homes and commercial buildings. When the company realized the excep-

Hearth of historic vertical lime kiln, Phenix Quarry.

Historic vertical lime kilns, Phenix Quarry (Greene County Historical Site).

tional quality of the limestone at the site, they stopped producing lime and began to produce and market marble.

State-of-the-art technology helped the company build its reputation. To power the massive machines for quarrying, cutting, and polishing the stone, a large powerhouse was completed in 1922 to house a stationary, coal-fired steam engine that generated electricity for the sprawling quarry complex and its adjacent company town. All this happened decades before electricity became generally available. The quarry had its own railroad tracks and a steam train for moving the massive blocks of quarried stone from place to place in the facility.

Flooded secondary pit at Phenix Quarry. Gang drills produced rectangular blocks which then were lifted out of the pit, reported to be some 75 feet deep.

Truly massive blocks of stone were cut using wire-line saws capable of cutting slabs 100 feet long and 8 feet high. Channeling machines cut deeply into the limestone, while gang drills made both vertical and horizontal cuts. The amount of hand labor required to keep the massive industrial facility in operation was enormous, yet the work force was so productive that even today the site is littered with hundreds of rectangular blocks of stone that weigh several tons each. So many, in fact, that the current owner used some of them to build a massive "fence" around part of the property.

The Great Depression ended the Cinderella run of Phenix Quarry. After the stock market crash, monumental buildings no longer were popular, so the great quarry lay idle for many years. The work force—from 250 to 400 men when production was at its peak—moved on, and the company town with its elegant buildings wasted away. Today there is but one small structure remaining as a reminder of what was once a busy, cosmopolitan town. The monolithic powerhouse that housed a huge stationary steam engine still stands in the quarry, but the engine was sold, shipped away, and reassembled on a new site. The homes of the workers were sold and hauled away; a three-room house could be had for $75, while a 5-room home brought $85.

Carthage Marble Company acquired the property and operated it for a number of years. The quarry is now locally owned and operated, processing waste rock into crushed stone for roads and construction aggregate, as well as agstone for farmers. The company also provides replacement marble for making repairs to buildings that used Phenix stone in their original construction. There is also a market for cut stone for modern buildings,

Weathered working face of almost chert-free Burlington Limestone. Smooth, even cuts were made by channeling machines.

such as the Center Church, about 4 miles south of the quarry. Among the artifacts of a distinguished past is a water well said to be 1800 feet deep. It was drilled to provide water for operating the steam power plant and to supply the water-using machinery used in quarrying, cutting, and polishing Napoleon Grey Marble. The well is 12 inches in diameter at the surface and one of the deepest wells in Greene County.

Phenix Quarry is a Greene County Historical Site, a veritable museum of bygone technology. But because the main pit is a working quarry, it is not open for casual visits. However, there is a second, much smaller pit just east of the main pit where one may see how Napoleon Grey Marble was actually quarried. In addition, two vertical lime kilns stand outside the active part of the quarry, and can be toured. They are constructed of massive stone blocks and have fireboxes lined with refractory bricks. One has a metal cover and smokestack still in place, but the supporting works for charging the furnaces with lime are no longer there.

NOTE: *The Phenix Quarry is a privately-owned, operating business and permission is required to visit any or all of the sites described here.*

Mineral Production Records

For many years, the U.S. Bureau of Mines collected and published mineral production statistics from the nation's mines, quarries, and wells, and made the information available for public use. But Congress abolished the agency in 1996, and transferred some of its powers and duties to the U.S. Geological Survey and various other agencies. However, because of a number of factors including confidentiality of company information, mineral production statistics are no longer readily available. Also, Missouri's State Mine Inspector collected and reported production statistics—particularly coal production—but that effort, too, is not as complete as one might wish.

Records from the 1800s and early 1900s are problematic in many cases. Lead and zinc production is variously reported as "xx carloads of ore," or "xx tons of product," which could mean raw ore, concentrated ore, smelted metal, or some other variant. When a dollar value is given, one has to know the price per pound, per ton, per short ton, or whatever unit is being used. Sometimes the records are completely pie-in-the-sky, as "The mine produced a large amount of ore."

CHAPTER 7

Sinkholes

Here in the Ozarks, the terms "sink" and "sinkhole" are used interchangeably, while a more complex assemblage of sinkholes is called a compound sinkhole.

Sinkholes and caves go together like cashews and chicken. Sinkholes show that soluble rock is being dissolved by groundwater, causing surface depressions while forming cave passages and rooms underground. Not all sinkholes, though, have underlying caves; some are simply caused by the slow dissolving of limestone bedrock that does not rise to the level of creating passages and rooms. Other sinkholes, usually associated with streets and roads, may be caused by broken water lines, defective drainage structures, or inadequate stormwater controls. Sinkholes tend to be more numerous in areas where the relief is fairly low, and the soluble rock—usually limestone or dolomite—is also the surface rock. This perfectly describes all of Springfield and much of Greene County, where the Burlington Limestone is exposed over 65 percent of the land area. Relief is at most a few hundred feet, with less-soluble rocks beneath the Burlington, a situation that favors springs, another positive geologic feature of the Greene County landscape.

Sequiota Spring at high flow; red color caused by a nearby sinkhole collapse that kept the spring muddy for weeks in December 2006 and January 2007.

Landscapes that form over areas underlain by soluble rocks are called karst, after places in the former nation of Yugoslavia where they were first studied. A complicated and unfamiliar terminology based on European feature names has developed and is used in the scientific literature, to describe sinkholes and related karst features. For example, a typical sinkhole is called a "doline," while a more complex, integrated sinkhole area is called a "uvula." Here in the Ozarks, the terms "sink" and "sinkhole" are used interchangeably, while a more complex assemblage of sinkholes is called a compound sinkhole. To be more precise, a sink would be a funnel- or saucer-shaped depression with no observable drainhole, while a sinkhole would look the same except there would be an obvious drainhole. Flying over a sinkhole area in a plane, the ground looks as though it had been cratered by bomb blasts. The Springfield area must have looked this way in the early 1800s, but as the city grew, some of the sinkholes were obliterated by filling, and structures were built over them. Today, when a sinkhole suddenly appears in the city, people are shocked, because there is no memory of what the landscape once looked like. People question, "How could this happen here?"

A classic example of this reaction occurred a few years ago in Nixa, just south of Springfield in a similar karst setting. A large sinkhole suddenly appeared in a residential area of small homes near the downtown area. Without warning, the sinkhole swallowed up a garage, complete with the homeowner's Chevy Cavalier automobile. The homeowner heard a noise, and when he went to check, there was a gaping hole where his garage and car had been. The hole was 125 feet deep, and at first the car could be seen in the bottom. However, the hole continued to grow until the car was covered, and it was not seen again. Nearby houses had to be evacuated, while residents checked their homeowners' insurance to see whether they had sinkhole coverage.

Close-up of dark red residual sediment from a sinkhole collapse in the recharge area for Sequiota Spring.

The sinkhole continued to grow for a few days, primarily by incremental subsidence around the sides. When it appeared stable enough to allow the use of machinery, the remainder of the damaged house was removed and a contractor filled the sinkhole with rock from a quarry. No attempt was made to recover the Cavalier; it remains—complete with gas in the tank—at the bottom of the sinkhole, likely in the recharge area for Blue Spring, on the south boundary of Greene County.

The Nixa sinkhole was a great shock to area residents, who had no idea they might be at risk of sinkhole collapse. But in 1974, a map showing the probable recharge area for Blue Spring showed a large number of coalescing sinkholes north of Nixa, leading to a swallow hole, a natural tunnel, and finally to Blue Spring. However, no dye traces were attempted during the time the Nixa sinkhole was open, so the linkage to

Sinkhole Map of Greene County

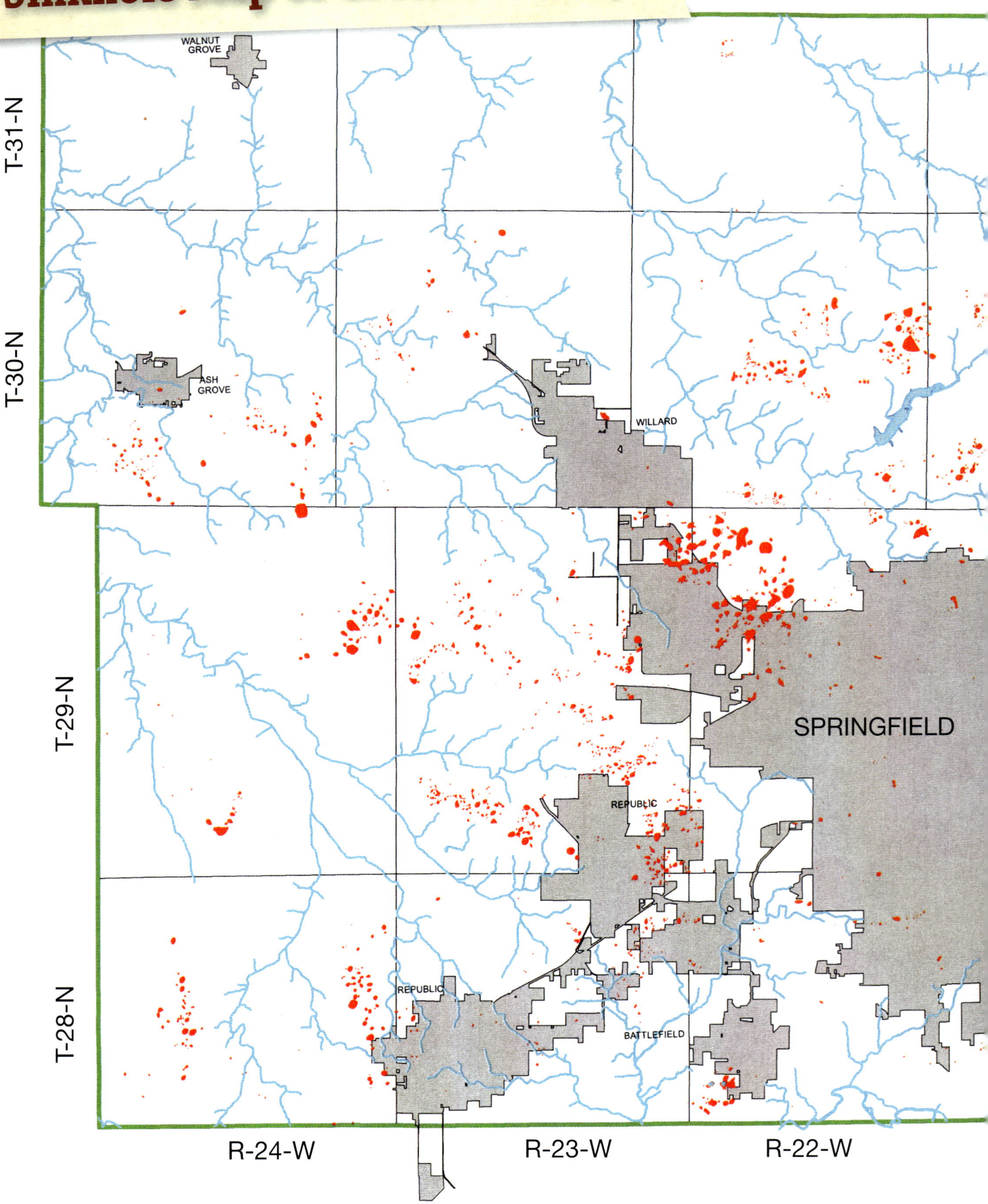

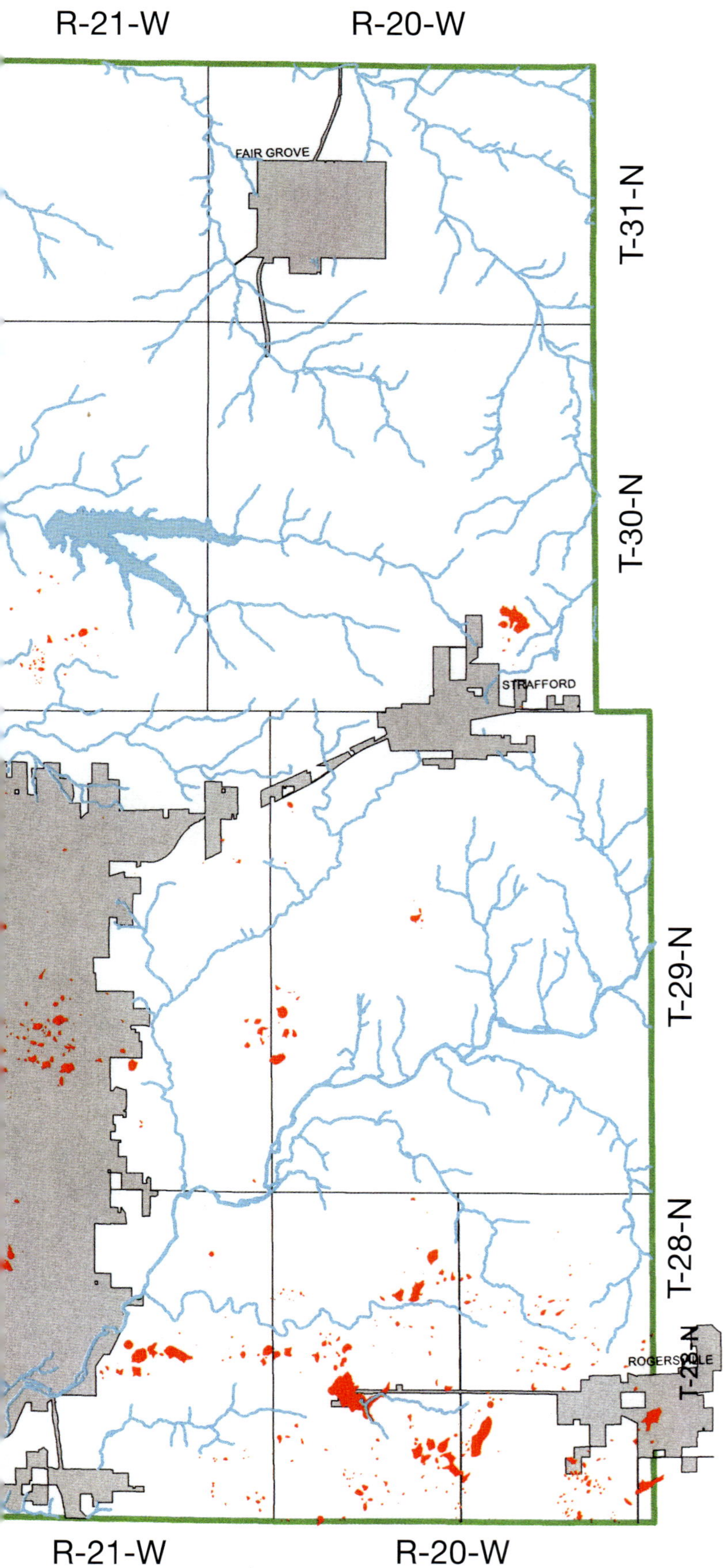

Sinkholes

Map compiled by Jim Vandiver, Greene County GIS

Blue Spring is speculation based only on topographic evidence.

Sinkhole collapses are so common in this area that both the city of Springfield and Greene County have developed regulations designed to minimize the negative impacts when sinkholes form incrementally or catastrophically. Over time, people and sinkholes have learned to coexist. Regulations ban building in or too near sinkholes, and numerous sinkholes have been modified to improve their beneficial functions, such as stormwater drainage and spring and aquifer recharge. There are also restrictions on filling sinkholes, because they are part of the natural drainage system that recharges groundwater aquifers and replenishes springs.

Sinkhole problems tend to worsen in weather extremes. In wet seasons, high runoff from rain or snowmelt causes more erosion, both on the surface and underground, sometimes triggering sinkhole activity. In dry years, the soil and surficial materials may dry up, crack, and lose cohesiveness, which may cause sinkholes to form. The year 2008, for example, was very wet and sinkhole events were common. Torrential rains along the Hwy. 60 corridor—an intense karst area with many sinkholes—flooded the westbound lanes of the road for more than a week when sinkhole water levels rose faster than the water could drain away. Rising water levels in the flooding sinkholes shut down traffic on a major east-west transportation artery.

While it may be beneficial to avoid changing the natural karst drainage systems, there are times when sinkholes must be filled or built over. When this happens, it is important to make provisions for drainage that replaces natural subsurface drainage with engineered solutions. In the 1960s, when Hwy. 65 was relocated to its present corridor east of Glenstone, a large sinkhole at the Cherry Street overpass had to be filled. The sinkhole held the entrance to Steury Cave, which could not be saved. MoDOT planned to fill the sinkhole and build the highway over it. However, the Missouri Geological Survey informed the highway designers that Steury Cave and its underground stream were essential parts of the Jones Spring underground watershed, and that filling the sinkhole and cave might interrupt the flow of Jones Spring and thereby negatively impact property values in one of the city's most attractive residential areas. In response, highway engineers designed a way to fill the sinkhole so that water could still flow beneath the highway, confined in a network of corrugated metal pipes (see illustration to the right).

Greene County has several reversible sinkholes (called estavelles, or resurgences) that are typical sinkholes in dry weather, but become springs in rainy weather. The Oval Sink, on the grounds of Springfield's Southwest Sewage Treatment Plant, is one of the best examples. There is another, Rader Resurgence, near Rader Spring on Wilson's Creek, and some others in the upper reaches of Wilson's Creek. Visualizing how these reversible sinkholes operate is easy if you think of them as overflow channels to be used in times when there is too much runoff for the underground drainage system to handle, so the overflow is vented through reversible sinkholes.

Over the years, area residents have learned to live with sinkholes. They complicate construction projects of all kinds, their unpredictability is legendary, and they tend to drive down property values. Still, sinkholes do have some positive attributes:

- **Natural—not constructed—stormwater drainage**. Sinkholes capture precipitation and release it gradually into the subsurface, thereby recharging groundwater aquifers.
- **Spring replenishment**. Sinkholes are the primary reason this area has so many springs. The sinkholes funnel runoff into the crevices and channels that comprise the supply networks that sustain springs through both drought and flood, which enhances property values and livability and attractiveness.

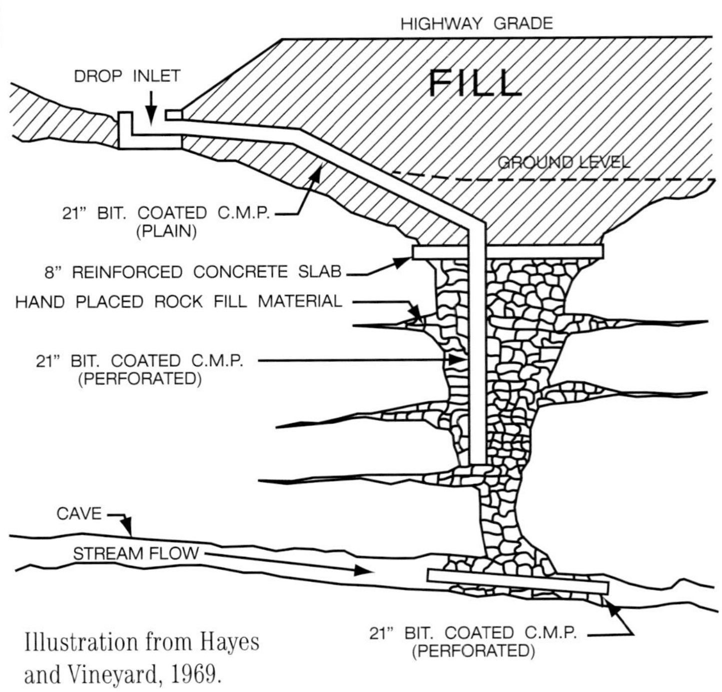

Illustration from Hayes and Vineyard, 1969.

- **Value-added golf courses**. Sinkhole-pockmarked tracts make good golf courses because their natural randomness promotes interesting course design, while the presence of sinkholes may mean the land is available at a lower price.
- **Event venues**. For years Springfield's signature Fourth of July celebration—Firefall—was held in a large, bowl-shaped sinkhole near the Springfield Airport. Firefall attracted tens of thousands of people. Though never proposed for the Guiness Book of Records, it may have held the most people ever assembled in a sinkhole at one time.
- **Law enforcement**. In 2008, a new sinkhole formed along the Galloway Creek Trail, at Covington and Lone Pine, on the southeast side of Springfield. It was a small feature, about 5 feet in diameter and 20-25 feet deep. No one paid much attention to it, until one day a fugitive being chased by police either climbed down or fell into the sinkhole, thereby eluding capture. But when he tried to climb out, it was impossible, so he had to yell for help. It was quite some time before a jogger on the trail happened to hear a faint voice coming from the sinkhole. He called 911, and when the police came, they got their man. As far as we know, this is the first time a sinkhole ever captured a fugitive and held him for the police.

There's an old saying, "If all you have is lemons, make lemonade." This is particularly apt in the Springfield area, where there is an amazing variety and number of sinkholes. Looking for positive uses for sinkholes rather than trying to eliminate them is far more rewarding than waging a costly and unproductive campaign to remove them from what is arguably an appealing and attractive landscape.

CHAPTER 8

Fourteen Places To See Geology

ONE

Badger Lead/Zinc Mine

What to see: Surface workings and tailings pile of the Badger Mine, which opened in 1906 or 1907, and operated intermittently into the 1920s.

How to get there: One of the few places to see evidence of early 20th century lead/zinc mining is the old Badger Mine, beside the swimming pool of the Cinnamon Square subdivision in east Springfield. From the Hwy. 65/Sunshine intersection, drive north on Eastwood to Cinnamon Place, to the pool. The tailings pile is a steep-sided mound about 25 to 30 feet high, on the northeast side of the pool.

Details: Standing in the parking lot of a modern community swimming pool, it's a stretch to imagine the scene more than a hundred years ago, when a shaft 285 feet deep penetrated an ore body that supported production of an estimated 75 tons of ore per day. The old shaft is now completely filled in, but concrete foundations of the ore processing building are still on-site. The old tailings pile—where waste rock was dumped—rises to the northeast. It is overgrown with gnarled trunks and twisty branches of Osage Orange trees, and laced with trails made by local kids.

The old Badger mine workings include several concrete bases for machinery, and a pile of tailings (waste rock), adjacent to a subdivision swimming pool. The mine shaft has been filled in.

In its heyday, the Badger Mine employed about 20 miners, who were paid $2.50 to $3.75 per day, good wages for the time. But it was dangerous work; one man died of bad air, and two others were crushed in 1907 when an ore hoist cable broke, dumping a load of ore on them.

The mine is on an extension of the northwest-trending fracture zone along Pearson Creek. The mineralized zone included the Northview Formation, with galena and sphalerite mineralization. Production records are limited, but the last operator of the mine, Daniel Mining Co., is said to have mined ore valued at $100,000. Competition from the Joplin District eventually shut down the mines in the Springfield area.

"The Mummy," a weathered gargoyle, Hwy. 65 South.

Grotesquely-shaped pinnacles of Burlington Limestone–here called "Gargoyles"–resemble crocs watching traffic pass on Hwy. 65 South.

TWO

Burlington Limestone "Gargoyles"

What to see: Rugged pinnacles of Burlington Limestone—here called "gargoyles" because of their fanciful shapes.

How to get there: Gargoyles are plentiful along the major highways in the Springfield area. Highway 65 from I-44 south toward Branson has arguably the best displays, especially those just north of the Battlefield Road intersection. I-44, James River Expressway, Hwy 13, and Hwy. 160 toward Willard are also well represented.

Details: The Burlington Limestone develops a very uneven upper surface, which is normally covered by residuum, soils, and vegetation, so there are few clues to the extreme unevenness until these materials are removed. Road cuts that expose the Burlington feature strangely-shaped, eroded rocks that are highly fossiliferous, hence the term, "gargoyles." The technical name for these features is pinnacles, and the residuum spaces between pinnacles are called cutters. In the Springfield area, highway builders have wisely left the gargoyles in place, which adds texture and interest to what otherwise would be bland, monotonous road cuts.

When the eroded Burlington surface is first exposed in construction projects, the fresh rock is tan to brown, while weathered sur-

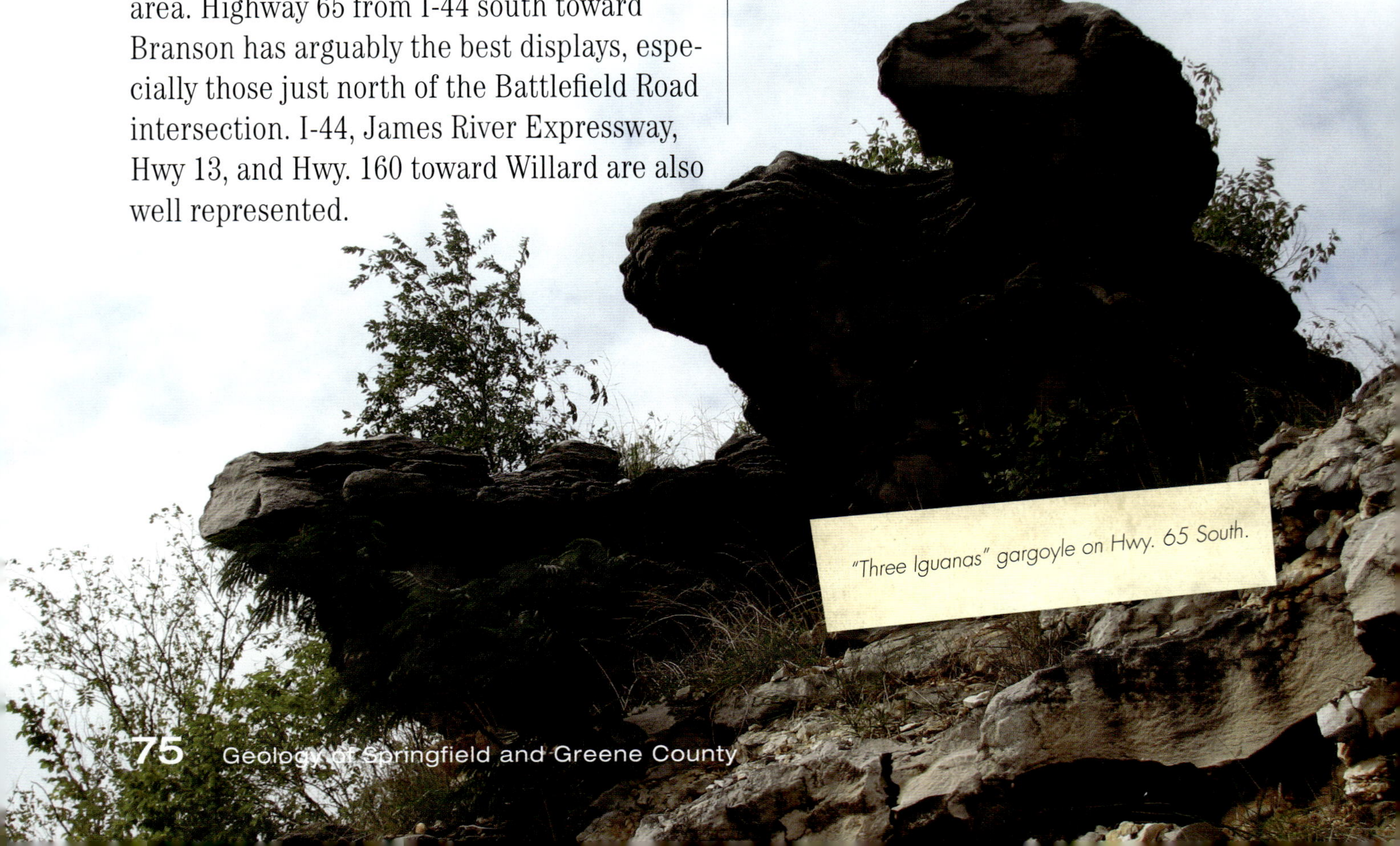

"Three Iguanas" gargoyle on Hwy. 65 South.

Close-up of a crinoid stem; gargoyles are composed of a "fossil hash," mostly of crinoid parts.

Close-up of crinoid parts, Burlington Limestone gargoyle.

faces are gray. However, after a few years exposure to sun and rain, the outcrops turn dark gray, sometimes almost black. The fresh rock rarely shows fossils, but as weathering proceeds, crinoids, brachiopods, horn corals and various other fossils emerge, showing that the Burlington is essentially a fossil "hash."

WARNING: *Gargoyles are mostly along major highways, so extreme caution is in order. If possible, use a frontage road; it may be hazardous to park on highway shoulders.*

THREE

Crighton Natural Bridge, Cave, Karst Canyon, And Spring

What to see: A karst complex with a natural bridge, a strong spring, a cave, and a scenic box canyon.

How to get there: From the intersection of Battlefield Road and Hwy. 65, drive east on Battlefield to a short street named Natural Bridge Lane. The karst complex is at the end of the street. Be aware that this area is residential and not open to public use. Visit only with prior permission.

Details: Crighton Natural Bridge is a segment of a former cave passage that looks much like a short tunnel about 15 feet wide and 10 feet high, with a length of about 50 feet. The bridge was named for Mr. Crighton, a former owner who modified it to make a clever storage building for his dairy business, in the days before refrigeration. He added wooden framing and closed the upper end with a door, put in a false floor to cover the rough cave floor,

Crighton Natural Bridge; note false floor and concrete door frames; door has rotted away.

Lower end of Crighton Natural Bridge as it opens into a karst canyon.

and built storage basins at the lower end in which to store milk.

The natural bridge, however, had no reliable supply of cool water, so Mr. Crighton built a flume from nearby Crighton Spring to keep the cooling basins filled. This was a highly innovative solution, because the flume had to be attached to the sheer face of a bluff of Burlington Limestone. The spring discharges from a small cave at a slightly higher elevation than the floor of the natural bridge, so the system provided constant flow to the cooling room, powered by gravity. The spring flows from the Burlington Limestone, at a constant water temperature of about 58 degrees. The flume—presumably made of wood—is long gone, but some of the steel bars driven into the limestone to hold the flume are still in place.

The natural bridge, cave and spring at the head of a very scenic box canyon make a tableaux that is a classic illustration of how karst features develop. There are several houses nearby that add a cultural dimension to the natural features. Mr. Crighton built a large three-story house of field stone that sits on the south side of the bridge area, and a large home of more recent construction is partly cantilevered over the north wall of the box canyon.

FOUR

Fair Grove And Fair Grove Mound

What to see: A combination of human and geologic history involving the old Wommack Mill and Fair Grove Mound, a monadnock (also called an outlier).

How to get there: The town of Fair Grove is on Hwy. 65, about 10 miles north-northeast of Springfield.

Details: Driving from Springfield to Fair Grove on Hwy. 65, one crosses the Eureka Springs Escarpment about 3 miles southwest of Fair Grove, thereby leaving the Springfield Plateau and entering the Salem Plateau. Approaching Fair Grove, Fair Grove Mound

Wommack Mill with sandstone "front porch."

Detail of Caudagalli (roostertail) fossils on a fieldstone in a masonry fireplace chimney.

can be seen on the right as a prominent wooded hill with communication towers and storage tanks of Greene Co. Public Water Supply Dist. No. 5 on its top. The Mound is actually a piece of the Springfield Plateau that survives as a remnant after long continued erosion. Such remnants are called "outliers" because they lie outside a major outcrop area.

The Mound is private property. An access road on the north end leads to the tallest tower and the water tanks, but permission would be needed to venture off the road. The best view of the entire mound is from Hwy. 125, which passes on the east side, about a mile away. To get closer, a farm road runs N-S on the east side, and another N-S farm road passes close to the Mound on the west side. Fair Grove Mound is capped by the Warner sandstone/conglomerate of Pennsylvanian age, which lies unconformably on strata of the Northview and Compton formations, of Mississippian age.

Fair Grove is a small town off the north end of Fair Grove Mound, known for its festivals centered on its mid-town park featuring the old Wommack Mill, built in 1883. Unlike most old Ozark water-powered mills, the Wommack Mill was steam-powered. Restored and maintained by the town, the mill sits beside a small stream that once provided water to fill the large concrete storage tank holding water for the boiler. The bedrock-floored stream exposes dolomite of the Jefferson City-Cotter formation.

The Wommack Mill has an unusual sandstone "front porch" built of massive blocks of sandstone, likely found locally. The porch

Fair Grove Mound in northeastern Greene County is a feature known as a Pennsylvanian outlier. That is, it lies outside the regional outcrop area of Pennsylvanian-age sediments. The top of the mound is sandstone and conglomerate, assigned to the Warner Formation.

Fieldstone chimney made of sandstone from the Northview Formation, on a historic cabin in Fair Grove.

is floored with smaller, flat stones—mostly sandstone—that tell of ancient, ripple-marked beaches and odd structures that may be traces of organisms that once populated seas in Ordovician times.

Just a few steps from Wommack Mill, on the hillside, stands a restored log cabin with a tall outside chimney that offers fascinating geologic stories. The chimney is built of stones that appear worm-eaten, as indeed geologists interpret them. Called "vermicular siltstone," they come from the Northview formation. Close examination of the chimney stones reveals several examples of strange fossils called "caudagalli" (literally, "rooster tails"). Paleontologists are not certain what organisms made these markings, but they are distinctive enough to be placed into two generic groups.

FIVE

Stonebridge (Mobley Natural Bridge)

What to see: A textbook natural bridge, with an associated cave, spring, and sinkhole, all in a small karst landscape.

How to get there: The site is in western Greene County, ON PRIVATE LAND, about 2.5 miles southwest of Bois d'Arc, on the north side of the Sac River.

Details: Stonebridge is the centerpiece of an Edenic setting in a small karst mini-canyon on the north bluffline of the Sac River. The name "Mobley Natural Bridge" comes from former landowners, though it might be more appropriately named after the current owners (for more than 35 years), because of their exemplary stewardship of the park-like area.

Stonebridge is formed in the Elsey Formation, a rock unit with alternating layers of limestone and chert, so the bridge looks as

Stonebridge looks as if it were a man-made bridge of masonry construction, but it is entirely natural, formed in the Elsey Formation, which is about 50-50 chert and limestone.

though it is man-made. The span is some 3 feet thick and 3 feet wide, with a flat top that is actually a bedding plane in the rock. The site is on the northeast corner of the intersection of the Sac River-Battlefield fault, which follows the Sac River, and the northeast-trending Sycamore fault. Some of the structural activity seems to be reflected in the northeast-trending karst mini-canyons hosting the natural bridge, associated spring, and overflow outlets for karst water during high-rainfall events.

"Persian Lion" pinnacle (gargoyle)

Stonebridge is a remnant of a former cave passage—now collapsed—that was formed by the spring, which is now at a lower level. There are two entrances to the cave. One is between the Bridge and the spring, giving access to a tight crawlway over gravel and cobbles. It is dry except in heavy rains, when flood flows pour out. The second entrance is in a small sinkhole on the hillside above the Bridge. A descending, crevice-like entrance gives way to a small passage leading to a room with ponded water. Beyond this pool, the cave appears to be blocked with collapse debris, apparently from another sinkhole farther up on the hilltop.

SIX

Pinnacle Forest At Riverbluff Cave

What to see: A "stone forest" of Burlington Limestone pinnacles (gargoyles) adjacent to the entrance of Riverbluff Cave.

How to get there: From the intersection of Republic Road and Cox Road in southwest Springfield, drive south on Cox Road toward Rivercut Golf Course and subdivision. Just before the Rivercut turnoff (Farm Road 190), a cluster of rock spires rises on a hillside on the west side of Cox Road.

Details: The erosion surface of the Burlington Limestone is noted for its irregular, pinnacled character, but rarely are the pinnacles so pronounced as they are in this "pinnacle forest." Some of the spires are up to 20 feet high, rem-

Burlington Limestone pinnacles resemble stumps in a cut-over forest.

Detail showing crinoid parts on a weathered surface

iniscent of the spectacular features of Bryce Canyon National Park, though on a much smaller scale. The Forest was uncovered during road construction that also led to the discovery of Riverbluff Cave, whose entrance is through a walk-in concrete entrance at the south end of the pinnacle grove. The Riverbluff Cave site and Missouri Institute of Natural Science are nearby, off FR 190. The Burlington is highly fossiliferous; crinoid columnals, stems, and calyxes are easy to find, along with horn corals, brachiopods, and other invertebrate fossils. Riverbluff Cave—which actually extends beneath the pinnacle forest—is well known for its vertebrate fossils including turtles, peccary, and mammoths.

SEVEN

Riverbluff Cave And Missouri Institute of Natural Science

What to see: The Missouri Institute of Natural Science displays a wide variety of natural history specimens, including fossils from Riverbluff Cave.

How to get there: From the intersection of Republic Road and Cox Road in southwest Springfield, drive south on Cox Road toward the Rivercut subdivision and golf course. Just past the Pinnacle Forest, turn right on Farm Road 190, then turn right again, toward the Institute buildings, on the hill to your right.

Details: Riverbluff Cave is an unprecedented discovery for southwest Missouri, an intact treasure trove of fossils revealed by blasting for highway construction, on the same day that terrorists flew jet planes filled with passengers into the twin towers of the World Trade Center in New York City. The significance of the cave was quickly recognized, and work began to secure the cave and analyze its remarkable contents.

Dr. Thomson was the first to comprehend the cave's potential value as a key to the past, but his life was cut short in a tragic accident before he could see the results of his work. At the Natural History Museum, which

Spectacularly preserved bear claw marks, believed to have been made by a species known as the Short-Faced Bear, which when standing could reach up to eight feet. Photo by Matt Forir.

After a construction blast opened an entrance to Riverbluff Cave, the first explorers saw striking speleothems in the first room, the prelude to discovery of trackways and fossil remains of Ice Age animals. Photo by Matt Forir.

Riverbluff Cave contains the only known tracks of Ice Age peccaries. Photo by Matt Forir.

includes classroom space, one may view fossils from Riverbluff Cave as well as many other natural history items that illuminate this region's rich geological history.

EIGHT

St. Luke's United Methodist Church (A Greene County Historical Site)

What to see: Fossiliferous fieldstone and cut marble in a historic landmark building.

How to get there: Take Hwy. I-44 west to Exit 67, then north on Hwy. T for a few miles to the village of Bois D'Arc. St. Luke's is on the south edge of town, visible for miles around.

Details: Approaching Bois D'Arc from the south, St. Luke's dominates the skyline. The structure is built of brown fieldstone, with thick mortar seams and extensive gray marble detailing on the front. Closer inspection reveals that the brown fieldstone is from local sources, highly fossiliferous, and multicolored in shades of tan to brown. Chert is the predominant rock type (likely from the Burlington Limestone), with sandstone and some burrowed siltstone from the Northview

St. Luke's Church at Bois D'Arc, a Greene Co. Historic Site. Church is built of fieldstone and cut marble, both from local sources.

formation. The individual stones are irregular in shape and size, fitted together like a giant jigsaw puzzle by skilled stonemasons.

The cut marble used for the front steps and detailing around the front door and front façade is likely from the Phenix Quarry, also a Greene County Historic Site. The marble [actually limestone of Mississippian age] used on the front steps is also highly fossiliferous, with crinoids, brachiopods and horn corals readily identifiable.

Fieldstone is mostly from the Northview formation, with many fossils, worm borings, and other sedimentary features.

NINE

Valley Water Mill

What to see: Sander Spring, three caves, sinkholes, a lake, a swallow hole, and an unusual geological structure called a horst.

How to get there: Going north on Hwy. 65, take the first exit past the I-44 interchange, and follow Valley Water Mill Road west for about a mile, to Valley Water Mill Park and the new Watershed Center.

Details: Valley Water Mill has a long and colorful history, and today it is a center for learning how to responsibly use and manage natural resources—especially water—in a sustainable manner. Five Learning Stations—Lakeside, Streamside, Springside, Wetlands, and Forestry—facilitate teaching about environmental stewardship. Hiking trails lead all the way around the lake, giving access to geologic features including springs, caves, rock outcrops, a horst, and a swallow hole.

The perimeter trail around the lake leads past Sander Spring, sinkholes, a horst, and two or three small caves. The horst appears to be a natural pass cut into the Burlington Limestone. In fact, there are two parallel faults cutting through the valley wall, with the space between them pushed

Sander Spring boils up from its submerged orifice in a Burlington Limestone bluff, at the south end of Valley Water Mill Lake. Potentially divable, the spring may flow from unexplored cave passages farther to the south.

upward. A similar feature—but with the space between the faults shoved downward—is called a graben. The flow of Sander Spring keeps Valley Water Mill lake filled. An old grist mill once stood near where the dam is today. The impounded lake water powered the mill through a traditional water wheel. Outflow from the lake enters Little Sac River and flows about a quarter mile downstream, to a swallow hole. There it goes underground and rises in Fulbright Spring, a part of Springfield's water supply system.

TEN

Stone Fence At Old Bluff School Site

What to see: The historic Bluff School site in west Springfield had a massive stone wall made of Mississippian chert fieldstone, topped with sharp spikes of chert to discourage student climbing. Many of the individual stones are fossiliferous and the construction work by WPA crews is unique.

How to get there: From the intersection of West Bypass and 160, go west to Farm Road 148, bear right on 148 to house number 4243. Stone fence surrounds a brown fieldstone house.

Gate post and part of perimeter wall for the historic Bluff School. Fieldstone from the Republic area, mostly chert and sandstone. The "spiky" upper surface was designed to deter students from climbing on the wall.

Details: The stone fence along 148 is about four feet high, with a prominent walk-in gate about midway. Near the western end of the fence there is a narrow opening for automobile access, built to accommodate Model T Fords and similar vehicles of the day. The fence was built in 1932 by the WPA [Works Progress Administration], created by the government to hasten recovery from the Great Depression. The stone fence completely surrounds about an acre of ground, on a low bluff along Wilson Creek, which adjoins the property on the west. The concrete foundation and front steps of the old, one-room Bluff School are still in place; the current owner built a home just west of the school foundation, but still within the stone fence. Building the fence took 25 men about a year to complete, using stones brought in from the Republic area on flat-bed Model T trucks. The fence features smooth, convex mortar joints—rarely seen today—and the individual stones with their many fossils make an interesting geological study.

ELEVEN

Lost Hill Park

What to see: The namesake Lost Hill, pinnacles in the bed of South Dry Sac River, six to eight caves, two natural arches, dairy farm building made of fossiliferous Northview vermicular [worm-eaten] siltstone, rock outcrops, and erosional landforms.

How to get there: From the east entrance to the Fairgrounds (Norton Rd. and Grant), drive north on Grant (becomes Farm Road 151) for about 2 miles, to the Park. Historic

Caudagalli ("rooster tail") fossils (center and right) on vermicular (worm-eaten) fieldstone, old farm building, Lost Hill Park

farm buildings including a stately residence, imposing red barn, silo, and a dairy building show the historical association of geology and agriculture. The Park is a waypoint on the developing Ozark Greenways Dry Sac Trail.

Details: Opened in late May 2010, Lost Hill Park is a geological gem. Ages ago, the South Dry Sac River flowed in a tight, sinuous loop to the north. As time passed and the loop tightened, the narrowest part of the meander loop was breached by the river, seeking to shorten its path. The river used a path pre-established by underlying cave passages. This shortening left behind a classic, teardrop-shaped Lost Hill, about 60 feet high and some 1900 feet long. Pieces of the cave system that originally caused the cutoff remain as small caves and two natural arches. The old farm building at the south end of the Park is built of fossiliferous, worm-burrowed siltstone of the Northview Formation. The stone was not quarried on the property, but came from a site near Noble Hill, farther north on Hwy. 13. The fieldstone walls of the

High Arch is high on a hill along the hiking trail through Lost Hill Park. It's counterpart–Low Arch–is at river level along the Little Sac River across from High Arch.

building are a veritable geology museum, with abundant worm borings and many examples of Caudagalli ("rooster tail") fossils. Actual outcrops of the Northview may be seen in roadcuts on Hwy. 13 north, near the Greene County line.

TWELVE

Crystal Cave

What to see: The second oldest tour cave in Missouri—open since 1893—is both a cultural landmark and a geological marvel. Visiting the cave is like time-traveling to an earlier generation, when life moved at a slower pace. The cave is graced by crystalline masses of dripstone and flowstone—hence the name, Crystal Cave—displayed in minimally-intrusive ways so the visitor can appreciate the cave for what it was—and still is.

How to get there: From the intersection of Glenstone and I-44, drive north on Glenstone (becomes Hwy. H) for 5 miles, to the cave. The entrance and parking lot will be on the left, marked by a sign. It may be helpful to call ahead, to make sure the cave is open for tours on that date.

Details: The human history of Crystal Cave is as intriguing as the geological story, though notably shorter. Alfred Mann, a cabinet-maker from Brighton, England, settled in Springfield in 1887, buying land known as "the old Jenkins place," on which there was a beautiful cave filled with crystalline stalactites, stalagmites, dripstone and flowstone. Mr. Mann and his family—including three teenage daughters—prepared the cave for public tours, building trails, clearing away breakdown, building steps and bridging crossings of the cave stream. The early venture in tourism opened in 1893, since compiling a 117-year history celebrating the underground beauty of caves. Alfred Mann's daughters—Ada, Agnes, and Margaret—devoted their lives to the cave, operating it until they passed away in the 1960s. Known simply as "the Mann Sisters of Crystal Cave," they never married, taking turns with cooking, housekeeping, and cave-guiding chores. In the 1800s, tours cost 10 cents, and were

Crystal Cave opened to the public in 1893, and has remained relatively unchanged for many decades. It remains one of Greene County's largest and most beautiful caves.

Detail of the left gatepost at Crystal Cave shows a section of stalactite (round stone) incorporated into the masonry. The other stones are mostly field stones from the Northview formation.

Fantastic Caverns is toured by Jeep-drawn trams, so it is uniquely accessible to anyone. This composite picture highlights the size and beauty of the cave and its many speleothems. Photo courtesy Fantastic Caverns.

conducted using kerosene lanterns. Electric lights were added much later, but as minimally as possible consistent with safety requirements of the State Mine Inspector.

Dr. Bretz wrote eight pages of descriptive text to explain the geology of the cave in his epic volume, "Caves of Missouri" (1956). To begin with, it is in an improbable host rock, the Elsey Formation. Caves are not common in the Elsey because it has so much chert. Stratigraphers estimate a 50/50 ratio of chert to limestone. Chert is silicon dioxide, a hard, brittle, nearly insoluble rock that occurs in lenses and/or nodules in a limestone matrix. As one tours the cave, chert is everywhere apparent. A feature named by the Mann Sisters is called "the upside-down well," because it is a ceiling cavity extending from the cave upward, leaving chert lenses in relief, where the soluble limestone has been dissolved away, leaving the chert behind.

Dr. Ken Thomson and students mapped the cave in detail in 1973. There is much to see and learn in Crystal Cave, but the marvel is simply that it has retained its mystique and its near-pristine character for well over a century, while providing income for its owners, geologic meaning and beauty to visitors, while contributing to the economic vitality of Greene County. May it ever be so.

THIRTEEN

Fantastic Caverns

What to See: Greene County's longest and arguably most beautiful cave, Fantastic Caverns is known far and wide as the cave you tour by Jeep-drawn trams. More than just a tour cave, Fantastic Caverns management teaches environmental responsibility through educational tours, workshops, and electronic media. The visitor center features educational exhibits as well as rocks, minerals and fossils. In addition to regular cave tours, visitors may hike a karst canyon nature trail through a picnic area adjacent to the Little Sac River, where Indian Spring, the resurgence of the stream in the Caverns, emerges at the base of an impressive bluff of Burlington Limestone.

How to get there: Public relations programs involving billboards, maps and brochures, and electronic media advertising have made the Caverns a well-known attraction. From the intersection of I-44 and Hwy. 13 in northwest Springfield, drive north on 13 about 2 miles, turn left on Fantastic Caverns Road, and follow "Caverns" signs about 3 miles to the visitor center and parking area. Guided tours by tram start at the visitor center; the hiking trail is self-guiding and accessible from the parking lot.

Details: Fantastic Caverns, discovered in 1862, has compiled an astonishing history under a series of ownerships and various names, yet today it remains a fragile, awe-inspiring place, accessible to almost everyone through the Jeep-drawn trams that travel a loop tour through the best-decorated parts of the cave. Through its long history, the cave has been a venue for cock-fighting, a speakeasy, a country music theater, and a "temple" of the Ku Klux Klan [The infamous KKK owned the cave for only a short time, in the 1930s, and lost it when they defaulted on a loan]. Considering its cultural history, it is amazing that it is still a "fantastic" place, graced by both massive and delicate mineral deposits—dripstone, flowstone, rimstone, helectites, and so on. The cave is regularly visited by speleology classes from Springfield universities, as well as science classes from area schools.

Fantastic Caverns is in the Burlington Limestone, a little over 2 miles long, so it is not even close to Missouri's longest caves. Yet its wide, spacious rooms and galleries are testimony to a complex cave-making history in which major cavities may have been dissolved by groundwater over a relatively long time period, followed by episodes of infilling by red clay and other sediment, then a much shorter period of secondary mineralization during which the impressive speleothemic displays grew—and are still growing.

Toward the far end of the cave, the Jeep road crosses a deep canyon, at the bottom of which a stream flows from a narrow,

Typical pile of rocks–mostly chert–from years of rockpicking by farmers to make the land easier to till and less wearing to machinery. In past years, farmers traditionally raided their rockpiles for materials to build farm structures and even homes.

Rock fence made of slabby Burlington field stones, on the north side of Rivercut Golf Course

low drain into the main cave, crosses the main passage, and exits through a continuation of the low drain passage. Exploration and dye tracing has shown that this stream comes from Smalley Sinkhole Cave.

Fossiliferous Burlington chert, in the Rivercut rock fence.

FOURTEEN

Rockworks

What to see: Rock houses, chimneys, rock fences, retaining walls, gateposts, stormwater control structures, spring houses, churches, cemeteries, farm buildings, and elegant, historic mansions abound in Springfield and Greene County. With a little familiarity with the rock types in this area, one can identify the source formations–such as the Burlington limestone, Northview burrowed stone, Roubidoux sandstone, and Jefferson City dolomite–study characteristic fossils, and appreciate the color and architectural diversity that geology brings to this region. Cemeteries have tombstones made of every imaginable rock type, usually from non-local sources, but some of the older cemeteries are more likely to have tombstones made locally, or simply use field stones.

How to get there: You can expect to find constructions made of rocks throughout Springfield and Greene County. While traveling the farm roads, look for houses and

Nathan Boone Home, with twin chimneys and foundation made of Northview fieldstone.

Detail of chimney masonry, showing colorful, burrowed Northview siltstone.

farm buildings made of stone, usually obtained locally. In the northern part of the county, rock piles are common in farm fields, the result of generations of rock-picking by farmers to make their fields more tillable and less destructive to machinery. In the past, it was common to use rocks from these piles to build homes and other rural buildings. When traveling anywhere in Greene County, keep an eye out for springhouses, which are as characteristic of rural areas as silos and weathered barns. In Springfield, there are hundreds of so-called "Ozark Rock Houses" that use native stone in distinctive structures that display geological features such as fossils, ripple marks, mud cracks, and mineral occurrences. Stone fences and retaining walls are also common throughout the City.

Fieldstones of Roubidoux sandstone, set on edge to form walls of Ozark rock buildings. Note wave ripple marks in the slab in top center. This type of rock masonry is sometimes called "giraffe rock" because it resembles a giraffe's coat.

Details: The Potter House and the Round Barn, both Greene Co. historic sites north of Springfield, are outstanding examples of the use of native stone. The Nathan Boone Home north of Ash Grove, the old mill at Fair Grove, the Methodist Church in Bois D'Arc, and the old farm building at Lost Hill Park also have fascinating rockwork worth a visit. In the City of Springfield, there are hundreds of rock structures ranging from simple retaining walls to stone fences, elaborate entry gatehouses, distinctive rock homes, and business buildings. The extensive rockwork walls and water control structures, along with the large rock swimming pool building at recently-renovated Fassnight Park are typical examples. The golf course at Rivercut, in SW Springfield, has a long stone fence on its northwest side, built of slabs of Burlington limestone and chert, resembling stone fences in Europe.

Rock houses in Springfield were documented in a detailed report by Ms. Debbie Sheals in 2005, titled "Ozark Rock Masonry in Springfield." The extensive report is an inventory of houses that comprise a genre of residences built of native stone, erected by stonemasons who were largely self-taught, using their own designs rather than following architectural plans. They used sandstone slabs set on edge, rubble stone, sawn sandstone resembling brick work, and field stone of several varieties. The builders mixed and matched designs and materials as they wished, resulting in some remarkable, eye-catching buildings that could be described as geologic folk art. The report includes street addresses, photos, and detailed descriptions for each property.

CHAPTER 9

Springs

As long as there have been springs, people have wondered where the water comes from.

Nameless roadside spring with water cress, northern Greene County.

The City of Springfield could easily have been named for the "field of springs" that existed when the first settlers arrived. Springfield's Park Central Square, on the National Register of Historic Places, has a water fountain "inspired by the waterfalls seeping out of limestone cliffs and bluffs around Springfield." Renowned landscape architect Lawrence Halprin, who passed away in 2009 at the age of 93, pioneered the concept of bringing green space, concrete, and water together in urban spaces. The "Natural Well" now marked with a plaque on Water Street was a window into a spring system that intrigued early residents. "Springs of Greene County," (2001, by Bullard, Thomson and Vandike) covers the history, geology, and hydrology of springs in considerable detail, building on the earlier work of Vineyard and Feder, "Springs of Missouri," (1974, 1982). Bullard, et al. document 210 springs in the city and county, plus an indefinite number of "ghost springs" that have been lost to urban development over the last century and a half. Balancing the lost springs, however, are the ephemeral springs, overflow outlets, and reversible sinkholes that swell the numbers after heavy and/or prolonged rainfall.

Conceding the losses over time, Springfield/Greene County springs remain cherished, honored, and ever-flowing natural resources. Fulbright Spring is an integral part of Springfield's water supply, while Sequiota, Ritter, Silver, Doling, and other springs are focal points of public parks. Many privately-owned springs, treasured by their owners, have been enhanced by landscaping, while others are much as they were in the early 1800s. In bygone years, spring houses were built over spring outlets for use as natural refrigeration. Some owners built elaborate and ingenious structures—like the bluff-hanging flume at Crighton Spring and Natural Bridge—to better utilize the relatively constant temperature and flow of natural springs. Old springhouses—like concrete farm silos—are vanishing icons of a way of life now forever changed.

Ray Springhouse, Wilson's Creek National Battlefield. Structure is made of limestone fieldstone, restored to its condition at the time of the battle.

Springs are natural outflows of groundwater. They range in size from small seeps such as the Ray Spring at Wilson's Creek National Battlefield, to huge, boiling rivers such as Big Spring at Van Buren, Missouri's largest spring. Greene County springs generally are smaller than their counterparts in central and eastern Missouri, but some—like the Winoka complex near the upper end of Lake Springfield—become raging torrents after heavy and/or prolonged rainfall. Spring water feels warmer in winter, cooler in summer, but the water temperature is relatively stable the year round. Springs are fairly evenly distributed across the county, with fewer springs in the northeast, where the bedrock (dolomite, sandstone, and shale) is less favorable for springs.

The general geologic setting is that a relatively thick unit of limestone—the Burlington-Keokuk—overlies about 65 percent of Greene County. The limestone is cut by fractures and joints, and its upper surface is rough and uneven, with a thick residuum of clay and chert, the products of long-term weathering. Because limestone is soluble in groundwater, the Burlington has become a classic karst topography with an extensive network of sinkholes, caves, springs, and losing streams. When it rains on karst topography, sinkholes act as natural funnels to channel the water into underground drainage networks, where it follows existing vertical fractures and natural partings (bedding planes) in the limestone bedrock. In limestone country, many springs are simply cave streams that flow out of cave entrances, as at Sequiota Park. Precipitation—rain, snow, sleet—is the ultimate source for spring water. While it may seem as if springs maintain constant flow throughout the year, measurements show that the flows closely follow precipitation, though the effects are damped by passage through soil, residuum, and rock permeability.

As long as there have been springs, people have wondered where the water comes from.

(continued on page 97)

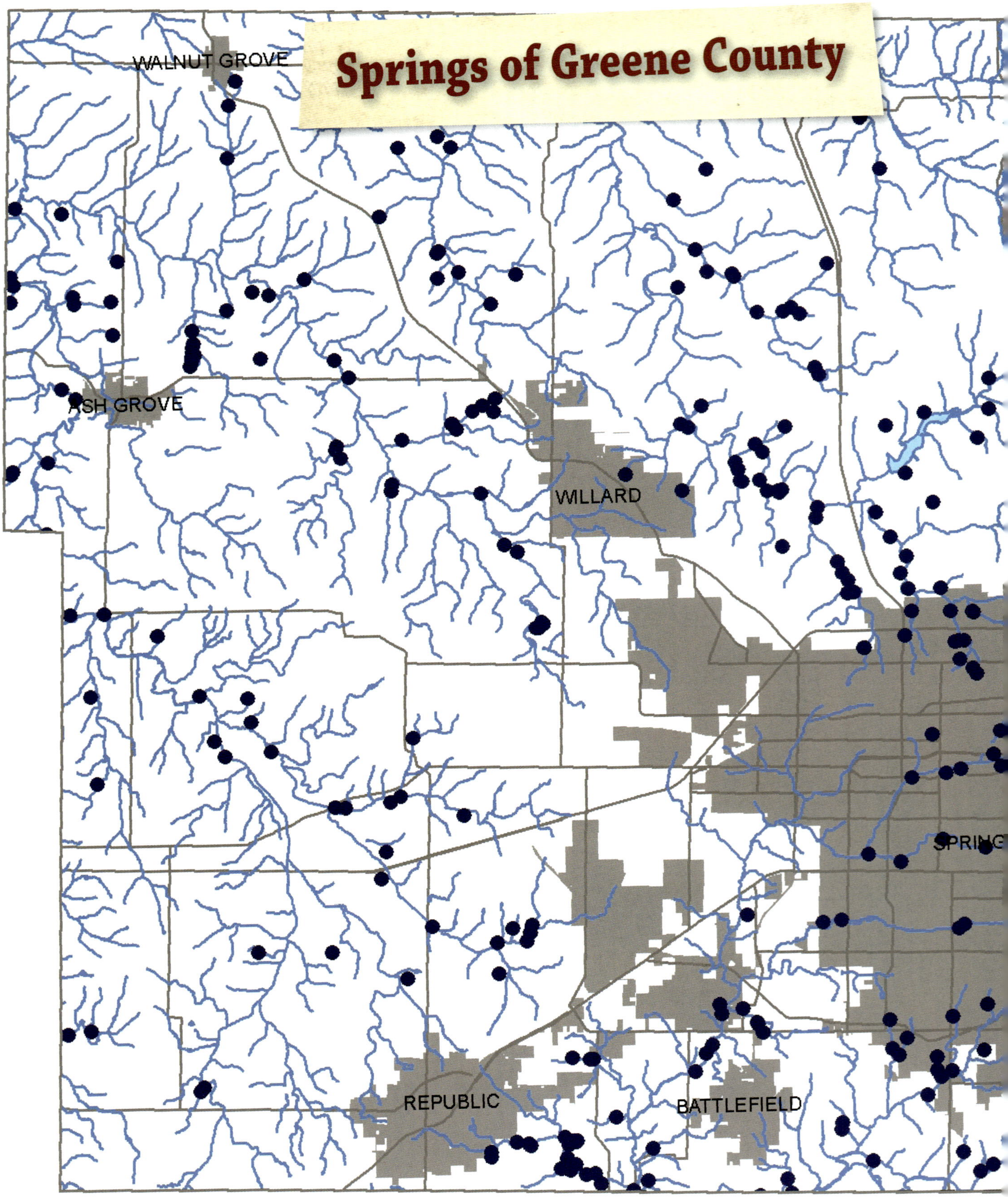
Springs of Greene County
WALNUT GROVE
ASH GROVE
WILLARD
SPRING
REPUBLIC
BATTLEFIELD

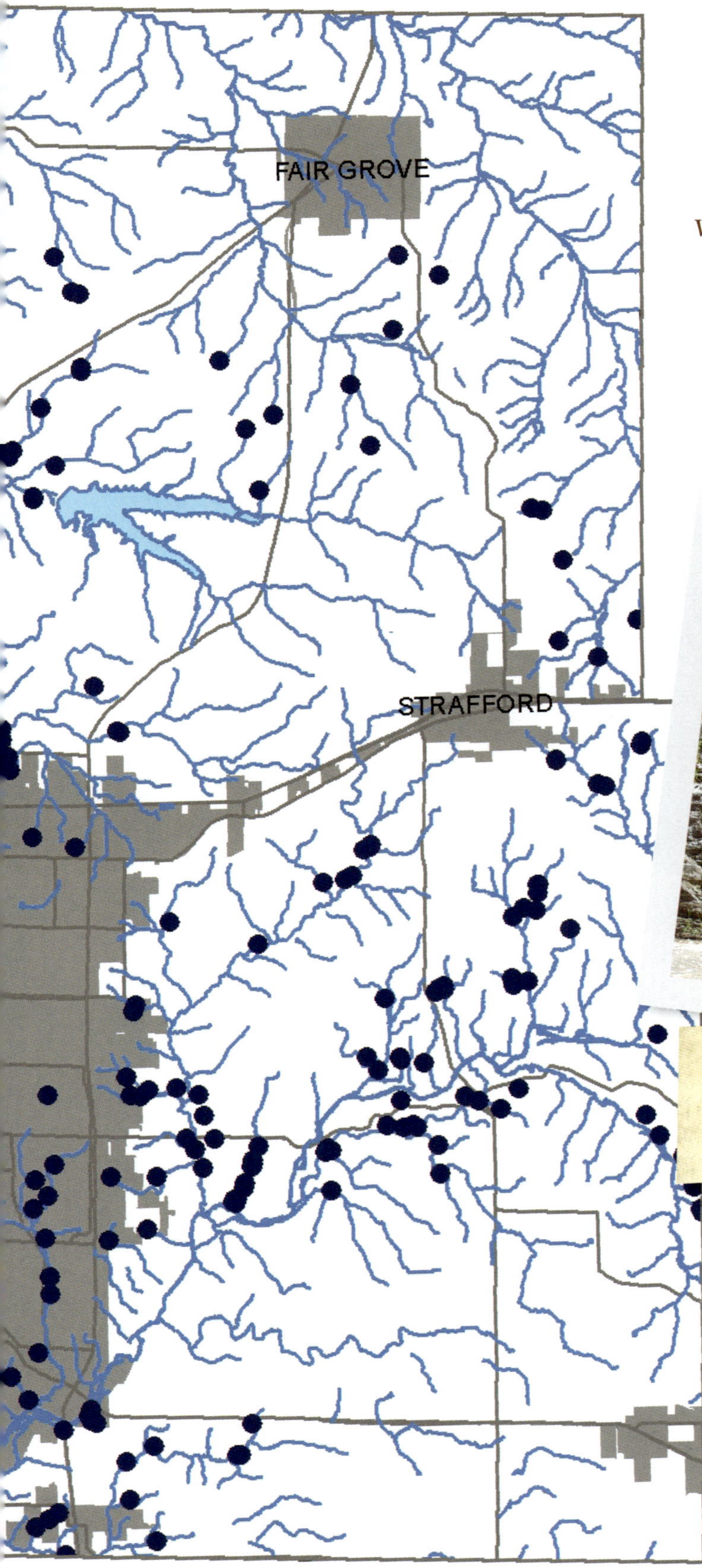

Fullbright Spring, one of the earliest water sources for Springfield, still contributes to the City's water supply. The spring is behind the metal grate on the lower right.

Map by Stacey Armstrong, Watershed Commitee

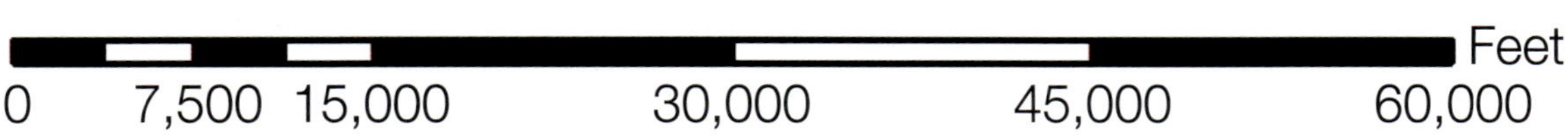

Jones Spring in normal flow.

Dr. E. M. Shepard pioneered early attempts to trace the movement of groundwater from losing streams, sinkholes, and swallow holes, to spring outlets at lower elevations. In the late 1960s, water tracing came into its own in studies by the Missouri Geological Survey (Vineyard, et al., 1967) in support of solving water pollution problems in Wilson's Creek, downstream from Springfield's Southwest Sewage Treatment Plant. This work, involving numerous traces using fluorescent dyes as tracing agents, defined flow paths from various sources that comprise a master spring system terminating in Rader Spring, one of Greene County's largest springs. While this study was underway, several unusual "reversible sinkholes" [now known as estavelles] were discovered. One of these is called the Oval Sink, at the lower end of the Southwest Treatment Plant property. Somehow, after extensive construction and plant improvements over a four-decade time span, the Oval Sink remains as it was when it was first described—a geologic transformer with a split personality: normally a placid, mysterious sinkhole, it becomes a powerful spring after heavy rains, pumping a flood of water into Wilson's Creek. Known to treatment plant employees as "the sinkhole that surcharges," the Oval Sink is a classic example of a geologic rarity.

Since the 1960s, other researchers have used fluorescent dyes to trace groundwater movement in various parts of Greene County. Not surprisingly, the directions of movement vary, and can be counterintuitive. In "Springs of Greene County," Bullard summarizes the body of work in considerable detail, revealing that while much is known, there is still a great deal to be learned about how Greene County springs operate.

When Hwy. 65 was relocated to Springfield's east side to relieve traffic congestion on Glenstone Ave., the new orientation passed through the underground drainage system feeding Jones Spring. A large cave known as Steury Cave was in the direct path of the proposed highway. The cave was to be dug open and filled, to allow the highway to pass over. Knowing the cave was tributary to Jones Spring, the centerpiece of one of the city's most exclusive residential areas, the Missouri Geological Survey pointed out that maintaining some semblance of natural flow through the interrupted underground drainage system would be imperative to avoid compromising the residential area. So the highway engineers came up with the solution shown in illustration on page 71. This was done, and Jones Spring continues to perform as designed, while the beauty and value of the residential area has increased.

Bullard, et al. (2001) classified Greene County springs into two major categories: Solutional (167) and Alluvial (36), with the remaining 7 in a miscellaneous group that includes seeps and other hard-to-classify features. The Solutional category includes springs that flow from cave openings, or seem to be associated with cavernous bedrock.

Sequiota Spring flows directly from a large cave entrance, and in fact can be explored by boat. Sander Spring completely fills its bedrock opening, while the Winoka spring complex issues from bedrock openings too small to be considered caves. Solutional springs usually occur in scenic settings, and historically they have been enhanced by landscaping and the addition of springhouses. In recent years, the availability of recirculating pumps

Oval Sink Estavelle in sinkhole mode.

Oval Sink, a reversible sinkhole or estavelle, in its outflow mode. Springfield Southwest Sewage Treatment Plant in the background.

Winoka Spring becomes a raging torrent after heavy rains.

has made it possible for "springs" to appear in unlikely places; not all flowing water is from a natural spring.

Alluvial (sometimes rendered "alluviated") springs rise through floodplain deposits along streams, or in other settings where bedrock is not apparent. Some of these may be associated with underlying fault zones, which may provide broken rock through which groundwater readily moves. Alluvial springs are likely to have more diffuse recharge, and may have more stable flow patterns than solutional springs of comparable size. Ranking Greene County springs by size is difficult, because there are so few discharge (flow) measurements. Many of the springs have no flow measurements at all, while others have non-specific notes as "low," "variable," "unk." [unknown], "seep," "mod" [moderate], or "high." Many others have even numbers (20, 50, 100, etc.), suggesting that the flows are estimated rather than measured. And there is no way to relate flow measurements to weather data. That is, a flow measurement made during late August–usually a dry, low-flow time in Missouri–is likely to be far lower than would be recorded during a normally wet season. Ideally, a spring would have numerous flow measurements spread over several years' time, and made at random intervals to insure coverage during a wide range of flow conditions. With this kind of data, a hydrograph can be constructed, that will show the flow characteristics of a spring so it can be compared with other springs.

While most of the springs in Greene County would be classed as "small," at least 20 discharge a minimum of 1 cfs [cubic foot per second], which is the standard measurement unit used by hydrologists and engineers. Based on the limited discharge data available, Rader Spring on Wilson's Creek is the largest spring in the county. When measured by the U.S. Geological Survey on April 28, 1965, its flow was 36.3 cfs, which equals 23,449,800 gallons per day. Only five flow measurements have been made on the spring, all during a two-year period in 1964-65. A new series of measurements may bring different results.

Springs rank high on anyone's list of most appreciated geologic features. They have a mystical quality that is appealing to everyone, they are colorful, scenic, thought-provoking, and various rare and/or endangered species find homes in their depths. They are here because the area's limestone bedrock is favorable for them to develop and function. They are also among the first geologic features to be harmed in pollution incidents such as spills, accidents, and inadequate environmental protection. Caring for springs—in all the ways the phrase suggests—should be a hallmark of this region's environmental policies.

CHAPTER 10

Caves

...the fact that caves and associated springs harbor rare and endangered species have made them even more significant public resources.

Cavers approach the cavernous entrance of Giboney (Doling Park) Cave. Springfield-Greene County Parks offers public tours of this cave to foster cave appreciation and understanding.

Springfield-Greene County has a splendid resource base of caves that bring an aura of mystery and the excitement of exploration and discovery to every generation. While they may not be the deepest, longest, or most beautiful caves in the world, they nevertheless offer a measure of all these elements, plus a rich history and a certain sense of the unknown that enriches life in southwest Missouri. From Crystal Cave that opened for candlelight tours in 1893 to Riverbluff Cave that literally exploded on the scene in 2001, and now brings real-time video to classrooms via fiber-optic cables, this area's caves are a distinctive and treasured part of the contemporary scene.

There are at least 25 caves in the Springfield-Greene County Parks system, including the well-known caves at Doling and Sequiota parks, which have been available for public access since the late 1800s. Parks has chosen to interpret caves as public resources worthy of responsible stewardship. They offer cave tours on a fairly regular basis, teaching appreciation for not only the earth science values of caves, but also their role in rare and endangered species management. The Missouri Department of Conservation owns Watkins (Speakeasy) Cave, on the Bois d'Arc Conservation Area in northwestern Greene County. The cave is gated, but available for touring by permit, and they hold "cave appreciation days" from time to time. Another cave education facility is available through the Springfield Plateau Grotto of the National Speleological Society. The group offers tours and training at a cave in Christian County, just barely outside Greene County. Training is offered in speleology and cave restoration, through award-winning programs developed by area cavers.

Caves are an emerging resource in Greene County. Despite having one of the nation's best-known tour caves in Fantastic Caverns, only about half of the county's 382 known caves have been surveyed, and many are not even fully explored. In 1956, when Dr. J Harlen Bretz's landmark volume, "Caves of Missouri," was published by the Missouri Geological Survey, fewer than a dozen caves were known in Greene County. Fifty years later, the number is approaching 400, an average of 5-8 new caves added per year. However, the discovery rate is not linear, but spikes in accord with cave exploration activity.

Dr. Ken Thomson, who joined the faculty of then-SMSU in the 1960s, immediately began studying caves, and with the aid of his students, documented and surveyed hundreds of caves across southwestern Missouri. In his native Utah, Dr. Thomson worked with underground mines, and he brought those skills to the inventory and analysis of Greene County caves. He taught his students how to map caves and pioneered one of the first college-level courses in speleology (the science of caves) in Missouri. Since his untimely passing in 2004, many of his former students have continued his dedicated study of cave resources to the point where caves are considered as significant resources rather than simply a curiosity or even a nuisance. Moreover, the fact that caves and associated springs harbor rare and endangered species have made them even more significant public resources.

Dr. Bretz actually saw only 7 of 12 Greene County caves he identified in his book. The others were undocumented beyond a simple verbal location or one given in land survey parlance. One–Jones Cave–is actually the outflow of Jones Spring in southeast Springfield, which has not been enterable because the spring fills the entire opening. The steady rise in the number of known caves is directly attributable to the persistence of volunteer cave-exploring organizations. Most Greene County caves are on private property, but caves are well represented on public lands and there are two well-known show caves–Crystal Cave and Fantastic Caverns–open to the public. Caves owned by Springfield

Intricately eroded wall rock in a Greene County cave, created when ground water dissolved limestone while the cave passage was completely flooded. Photo by Roy Gold.

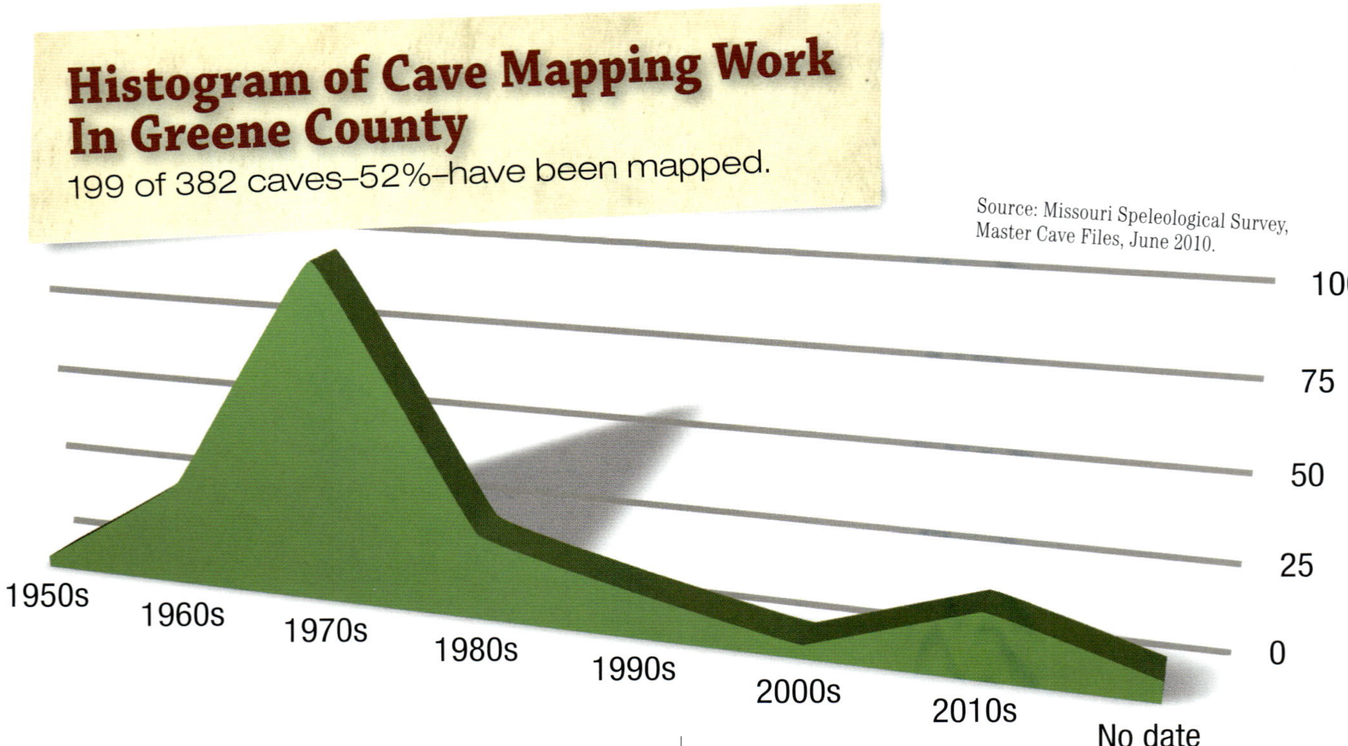

Parks are available to visit, subject to various restrictions based on such considerations as whether or not endangered species may be present, flood risk, or other sensitive conditions. Because a large percentage of caves in Greene County are vulnerable to flash flooding, it is always prudent to check weather before entering any cave that may flood.

The number of caves in the county changes frequently. For example, Riverbluff Cave–the most spectacular new cave discovery in modern local history–suddenly appeared when an explosive charge being used in road construction blasted an entrance into a hitherto unknown cave. There's an old saying, "Truth is stranger than fiction," demonstrated to the nth degree in the story of Riverbluff Cave. Greene County road crews were extending and widening Cox Road to the south, to the James River and the Rivercut development. They needed to make a cut through Burlington Limestone, so the rock had to be drilled and blasted away. All was going according to plan when a dark opening appeared after a blast, something that was totally unexpected. And it likely would have been dealt with in another series of blasts had it not been September llth, the infamous date of the bombing of the twin towers of the International Trade Center in New York City 9-11. All blasting was halted immediately, so Dr. Ken Thomson and county officials had an opportunity to look into the dark opening. They found themselves in a large, well-decorated cavern with pristine passages leading into darkness. Quickly Thomson and his party realized this was no ordinary cave, but a treasure trove of Ice Age paleontological riches.

Plans were changed to avoid any further damage to the cave, and the accidental entrance area was stabilized and fitted with a manhole-type artificial entrance. However, before the cave could be totally secured, vandals broke in and caused considerable damage, mostly to the cave's striking mineral formations. Fortunately, the vandals were caught, the damage was repaired by volunteer cave restoration specialists, and the perpetrators were punished through the court system.

View down a tubular passage in Ash Grove Cave, western Greene County, typical of cave passages in the Burlington Limestone. Note cave fill and breakdown blocks. Photo by Charley Young.

Today Riverbluff Cave has a new Natural History Museum on the site, designed for the study and interpretation of a cave now recognized as one of the oldest Ice Age paleontological sites in the nation, having fossil vertebrate remains dating back more than 1.1 million years. Megafauna such as wooly mammoth, short-faced bear, and camel share the spotlight with peccary, various snakes and turtles, along with numerous microfaunal species. The cave even has fiber-optic cables installed to facilitate real-time video to link classrooms with researchers working underground.

Riverbluff Cave and its remarkable contents are expected to draw international attention and full scientific investigations are expected to continue for many years as the cave slowly reveals its complex and fascinating history to meticulous researchers.

Cave Distribution

Greene County caves are scattered diagonally across the county in a northwest-southeast swath, approximately the outcrop area of the Burlington Limestone. Because most caves are privately owned, in respect for property rights specific locations are not given. By far the largest numbers of caves occur in the Burlington-Keokuk Limestone, which underlies some 65 percent of the county land area. A few caves are found in the Pierson formation, and a few others–notably Crystal Cave just north of Springfield–are in the Elsey formation. Limestone of the Elsey is fine-grained and favorable for cave development, but the formation is approximately 50% chert, which greatly limits cave-making. The geology in northeastern and southwestern Greene County is much less favorable for cavern development, so few caves are found there.

Compared with the 80 cave-containing counties in Missouri, Greene County stands number three on the list, behind Perry (629), and Shannon (403). As of June 2010, Greene County had 382 caves. Counties surrounding Greene also have considerable numbers of caves: Christian 225, Stone 306, Taney 157, Webster 84, Dade 55, Polk 38, Dallas 27, and Lawrence County 44. Greene County caves, while numerous, are not notably large or deep. Roughly half of the mapped caves are less than 100 feet long, about a third are from 100 to 1,000 feet long, and the remainder are

Caving Technology

Caving today is a far cry from what it was 50 years ago when caving helmets were often home-made, or adapted from mining equipment. The best lights were smelly, quirky carbide miners lamps, and compasses for mapping were frequently military surplus instruments. Serviceable surveying tapes, notebooks, clinometers and drafting equipment usually were adapted from whatever could be found on the market. Now, caving helmets and other gear are specially designed and readily available through dedicated suppliers. Lighting has embraced LED technology, wet suits make caving in water crawls bearable, and waterproof mapping compasses and clinometers make mapping easier, while laser distance-measuring devices and waterproof notepaper simplifies in-cave surveying work. When the field work is done, computer programs can quickly transform notes and sketches into precise, professional-looking cave maps without the laborious technical drafting once required to complete a cave map. Caving gear now routinely includes diver wet suits and sometimes SCUBA gear for long water crawls and flooded sumps that once would have halted exploration. On the surface, GPS units now deliver precise locations and the internet offers a bewildering array of maps, aerial photography, and space imagery along with digital elevation models, making child's play of once-difficult spatial relationships. Unfortunately, though, the human aging process still cannot be circumvented.

longer than 1,000 feet. Fantastic Caverns is currently the county's longest cave at 6,823 feet, or about one and a quarter miles long. Some 50 caves have been lost over time to highway construction and urban growth. These caves have been filled and/or obliterated, or otherwise made inaccessible. The loss of cave resources to urban growth is likely to continue, perhaps at a much slower rate as the natural resource value of caves continues to rise, spurred in part by their role as refugia for endangered species. On the other hand, new discoveries made during excavations as well as natural sinkhole collapses will result in new caves added to the list of significant caves.

Caves develop primarily in limestone or dolomite, because these rocks are soluble in groundwater. Limestone is usually more soluble than dolomite, so there are far more caves in the limestone outcrop areas than in the dolomites. Most are in the Burlington Limestone, with a few caves in the Pierson Formation, and even fewer in the Elsey Formation (Crystal Cave is a notable exception). Cave-making occurs in rock strata younger than the Northview formation, which limits downward solution because it has a significant shale unit that is not soluble in groundwater. The deepest known cave in Greene County is about 85 feet deep. There are no deep caves or springs in the county. This is not the case elsewhere in southwest Missouri, where the Northview is breached, notably in Marvel Cave at Silver Dollar City. This makes possible a deep cave beginning in Mississippian Limestone, cutting through the Northview shale in a narrow, perhaps fault-determined passage, and extending far down into the underlying Ordovician strata.

Cave Entrances

By far the most common entrance type is the outflow entrance, typified by Sequiota Cave and Giboney (Doling Park) Cave. Usually outflow caves have entrances in bluffs or hillsides, with a cave stream or spring flowing out. The entrance area will usually be the lowest point in the cave. Following the water upstream leads up-gradient toward the end of the cave, or at least the limit of human exploration. Older outflow caves—usually at higher elevations in bluffs—will be dry, with only intermittent flow after heavy rains. Most outflow caves have only one entrance, but some may have multiple

Massive stalagmitic speleothems in Fantastic Caverns

Actively growing flowstone filling a small side passage in Giboney Cave, Doling Park.

entrances, either outflow or inflow (pit or shaft) entrances. Others may have artificial and/or hard to classify entrances. Natural tunnels, of course, have both an inflow and outflow entrance, and may even have additional "skylight" entrances if there has been roof failure.

Pit or sinkhole entrances are more common in upland areas. They usually form when cave roof failure results in a sinkhole on the surface. Or, in the case of Riverbluff Cave, blasting for road construction revealed a cave passage. Gravelle Pit Cave, in east Springfield off Sunshine, is a pit-entrance cave that was unknown until a sudden collapse revealed a 76-foot shaft. Crystal Cave, the venerable show cave north of Springfield off Hwy. H, also has a sinkhole entrance on the side of a hill.

Cave Length

The longest known cave in the world is the Mammoth/Flint Ridge Cave System in western Kentucky, in Mammoth Cave National Park. It currently has about 400 miles of underground passages, and continues to grow as exploration continues. Wind Cave and Jewel Cave–complex three-dimensional mazes in South Dakota's Black Hills–each have more than a hundred miles of twisty, interlinked passages. The longest cave in Missouri–Crevice Cave in Perry County–is over 28 miles long and it, too, continues to grow. Compared to these monster cave systems, Greene County caves seem insignificant in comparison.

Fantastic Caverns sits atop the county's long cave list at 6,823 feet, roughly 1 ¼ miles long. This seems insignificant, matched with far longer caves, yet since its commercialization (then called Temple Cave) in 1895 Fantastic Caverns has delighted many thousands of visitors and continues to do so in the 21st century. Its beautiful speleothems—carefully protected from vandalism—are a delight to the eye and as "fantastic" as they ever were. Tiny blind, white cavefish still live in the cave stream, proving that it is possible to maintain and enjoy the natural world in a technological era.

The long cave list currently has 22 caves longer than 1000 feet. If all those caves could be lined up end-to-end, their lengths would total about 12 miles, yet the value of these natural features to the Springfield area is incalculable. Consider the long-term economic impact of Crystal Cave, Fantastic Caverns, Giboney Cave, and Sequiota Cave, all of which date from the late 1800s, and are still productive. Cave length is an imperfect measure of the worth of a cave. Riverbluff Cave is just over

Caves, Cavers, and Caving

Caves are an ever-fascinating part of our landscape and culture, in large part due to the writings of Mark Twain, whose fictional characters Tom Sawyer, Huckleberry Finn, and Becky Thatcher made cave exploring—or caving—a rite of passage for youngsters in areas that have cave resources. Luella Agnes Owen wrote about Missouri caves in her 1898 book, "Cave Regions of the Ozarks and Black Hills." She showed by her own cave-exploring exploits that female explorers were as capable as males. Inexplicably, Miss Owen passed through on her way to Marble (now Marvel) Cave, without visiting any caves in Greene County. She was a champion of careful, responsible stewardship of caves, a pioneer environmentalist. She famously opined that people should visit caves in the Branson area "...before the railroad gets there."

"Cavers" are people who explore, map, and describe caves, and "caving" is what they do. Mostly young, dedicated people in their teens, 20s, and 30s, cavers organize themselves in clubs—called grottos—and spend time exploring, mapping, and studying caves. Without cavers, we would know little about Missouri's outstanding cave resources. Caving became a force in resource management only in the past 50 years, brought on by the publication of the book, "Caves of Missouri," in 1956. The book spurred the organization of a consortium of cave-exploring clubs named the "Missouri Speleological Survey, Inc." (MSS). Dr. Thomson was a strong supporter of the MSS, serving as its President for 13 years.

Jon Beard and Chris Gertson of Springfield Plateau Grotto restore a stalagmite. Photo courtesy of Springfield Plateau Grotto.

Luella Owen used candles and wore a split skirt for cave exploring. Dr. Bretz carried a flashlight and Coleman lantern, while exploring caves with his collie dog, Larry. Today's cavers use LED lighting technology, purpose-designed caving clothing, diver wet suits for cold, watery passages, laser distance-measuring devices, and waterproof compasses for mapping. GPS units pinpoint cave locations and guide repeat visits to obscure cave entrances. Even so, only about half of Greene County's known caves have been thoroughly explored and documented. However, a coordinated effort is now underway to complete the documentation of the county's cave resources by July 2015.

Despite having a reputation as being a cold, wet, muddy and dark pastime, caving generates

(continued on page 109)

Cave formations (speleothems) broken by vandals can be restored using epoxy cement made in Springfield by the 3M company. Jon Beard spreads epoxy on a broken stalagmite. Photo courtesy of Springfield Plateau Grotto.

Caves, Cavers, and Caving (continued)
a strong conservation ethic in its practitioners, reflected in their credo, "Take nothing but pictures, leave nothing but footprints, and kill nothing but time." Greene County cavers have developed a national reputation for their efforts in cave restoration. Restoring broken speleothems using high-tech epoxy adhesives, erasing graffiti using non-destructive methods, removing trash and offering training in responsible caving help insure that Greene County caves will be treasured, not obliterated.

a quarter mile long, yet it contains strikingly beautiful speleothems and sufficient paleontological attributes to keep researchers occupied for decades. When the science is done, the cave will still be there, enriched by history unimagined when people first saw what lay underground. The Greene County Long Cave List is certain to grow in the years ahead. Ash Grove Cave—named for the town in northwestern Greene County where it is located—was first mapped in 1969, at 4,555 feet. Remapping is in progress, to take advantage of advances in exploration and mapping technology, and the footage already exceeds that of 1969. As time passes, new caves will be added to the list, and caves now on the list will gain new stature as cavers continue to push passages once thought to be too low or too wet or too tight. So the list will grow both in numbers of caves and their lengths, but the geological limitations imposed by stratigraphy and relief insure that Greene County will never have record-breaking long or deep caves. That's the way it is.

Cave Patterns

Greene County caves are usually simple in pattern, with a main passage that may have one or more side passages. Some caves in plan view resemble the branches of a tree, and so are said to have a dendritic pattern, while others may have a zig-zag or en echelon pattern. There are also a few caves with more complex–even mazelike–passages. The orientation of straight passages or parts of a cave is likely caused by joints, fractures, or faults in the bedrock, while amorphous patterns may reflect bedding plane control.

Passage Shapes

As a general rule, caves in Greene County are likely to be low, wet, and muddy, so cavers anticipate those conditions and dress appropriately. Fortunately, there are many exceptions to that presumption, that make caves exciting and challenging to explore. Big Cave on the Sac River in western Greene County

Long Caves of Greene County

Source: Springfield Plateau Grotto, the National Speleological Society, 2009.

Name	Length (in feet, m = measured, e = estimated
1. Fantastic Caverns	6,823 m
2. Ash Grove Cave (1969 map; resurvey underway)	5,280 m
3. Big Cave (cave only partially surveyed)	5,280 e
4. Keeslar Cave (1966 map; scheduled for resurvey)	4,000m
5. Kelly Cave	4,000 m
6. Crystal Cave (some unsurveyed leads)	3,610 m
7. Sequiota Cave	3,500 m
8. Smalley Sinkhole Cave (hydrologically connected to No. 1)	3,483 m
9. Steury Cave (mapped in 1968, but entrance filled)	2,384 m
10. Swallow Hole Cave	2,000 m
11. Low Water Bridge Cave	2,000 m
12. Siphon Spring Cave	2,000 m
13. Hall Branch Cave (map in progress)	2,000 e
14. Craig Cave No. 2	1,898 m
15. Riverbluff Cave	1,885 m
16. Watkins Cave	1,750 m
17. Giboney Cave (1974 map; resurvey in progress)	1,500 e
18. Bedding-Joint Cave	1,500 e
19. Scott Spring Cave	1,500 e
20. Sams Cave (map in progress)	1,200 e
21. Williams Cave System	1,100 m
22. Treasure Cave	1,000 m

is thought to be the county's longest cave, yet its entrance is so low and so wet that few people have gone into the cave more than a short distance. It is not unusual to encounter crawlway passages connecting other parts of a cave that have large, pipe-like (sometimes called borehole) passages, and/or big rooms. In general, passages controlled primarily by the separations (called bedding planes) between layers of bedrock are low and wide. In contrast, passages controlled by fractures or joints are likely to be high and narrow, and apt to run in straight lines. A large cave may have a combination of passage shapes. Groundwater moving vertically—especially in the Burlington Limestone—frequently

Jones Spring issues from a flooded cave entrance and cascades down Burlington Limestone.

dissolves pits or shafts that add dimensional variety to caves. When such features are open to the surface–as when a sinkhole collapses–they are called pit or shaft entrances. When one encounters such a feature from the lower end, as when exploring, it is called a dome or domepit. There are many of these in Greene County, but because they are usually limited to the Burlington Limestone, the deepest known shaft/dome is about 85 feet.

Speleothems

The word "cave" brings to mind spectacular displays of "cave formations," including stalactites, stalagmites, and flowstone. These displays are–or once were–present in nearly every cave, but sadly, thoughtless visitors have broken a lot of them. The word "speleothem" covers all kinds of secondary mineral deposits in caves, including all of the common variations of dripstone and flowstone, rimstone, anthodites, cave pearls, cave coral, moonmilk, gypsum flowers, and hundreds of other variations. These mineral displays form in caves after the openings are already there, so they are called "secondary" mineral accumulations.

Highly decorated caves usually have extensive displays of minerals formed by dripping or flowing water, so they are collectively called "dripstone" or "flowstone." More often than not, the two variations occur together and morph seamlessly into each other. "Forests" of small stalactites called "soda straws" are commonly found growing from cave ceilings, with corresponding stump-like stalagmitic growths on the cave floor. Exotic forms such as anthodites, spathites, cave coral, and moonmilk are rare, but not unexpected.

Compared with the world's great caves, Greene County caves may seem insignificant, but they have many of the same attributes that make Mammoth Cave, Carlsbad Caverns, and Blanchard Springs Caverns destination places. Best of all, they are right here, underfoot!

CHAPTER 10

Near Miss From Outer Space

The impact must have caused an incredibly powerful explosion, capable of destroying everything nearby and for tens of miles away.

The year was 1949. A young geologist named Thomas R. Beveridge was mapping the geology in the vicinity of the village of Weaubleau, about 50 miles north of Springfield. It was a quiet, rural area, and Beveridge enjoyed the solitude as he measured and recorded the rock strata in his field book. He carried the geologist's standard kit of tools: a well-worn geology pick, hand lens, Brunton compass, hand level, and a small bottle of dilute hydrochloric acid (used to differentiate between limestone and dolomite). The base map he used was a standard topographic quadrangle map.

World War II had ended in 1946. Beveridge's career had been put on hold while he was in the Army Air Corps, serving as a tail gunner on a B-24 Liberator bomber. On a mission to bomb the Ploesti oil fields in Romania, Tom's plane had been shot down. But he survived, and following the war he completed his PhD degree from the Univ. of Iowa, and began his geological career with the Missouri Geological Survey and Water Resources in Rolla. Later he would become one of Missouri's best-known State Geologists, but for now, he had had enough excitement; instead of simply trying to survive the War, he was content to pit his education and training against the unusual geology he found in the hills and hollows of the Weaubleau-Osceola area.

One day he found some rock exposures that seemed out of place, chaotic, and altogether different from the rather straightforward sedimentary geology he was accustomed to finding. Older rocks seemed to have been violently thrust into younger strata, and he noted a lot of uncharacteristic broken rock called breccia, that was not present elsewhere. He suspected thrust faulting, but he could not make sense out of the spatial arrangements of the rock masses that he found. Nevertheless, Beveridge recorded his findings and called the area the "Weaubleau Structure" without offering a definitive explanation of how it had originated.

The Survey published Beveridge's findings in 1951, in a report titled, "Geology of the Weableau Creek Area." In the report, he considered several possible origins for what he had found, but felt that he did not have enough information to make a definitive conclusion, so he termed it a "Cryptoexplosive Structure," meaning a disturbed area that had undergone some sort of explosive activity, but the cause could not be determined.

Beveridge noted that his "mystery structure' happened to lie along an axis of similar places that geologists had called the "38th Parallel Lineament" because it roughly coincided with the 38th parallel of latitude. Beginning with the Rose Dome feature in Kansas, the alignment continued eastward with the Weaubleau structure, the Hazelgreen disturbed area along I-44, the Decaturville Dome south of Camdenton, the Crooked Creek complex near Salem, the Furnace Creek volcanic feature near Caledonia, the Avon Diatremes of Ste. Genevieve County, and ending with the Hicks Dome in Kentucky. The line of disturbance suggested a zone of activity caused by some underlying fault or zone of failed volcanic events. The disturbances all had different characteristics and some—like the Hazelgreen and Furnace Creek features—were known only from drillcore obtained by mining companies exploring for mineral deposits. At the time, the Viburnum Trend (also called the New Lead Belt) was in development and mining geologists were doing extensive exploration drilling to define new heavy-metal deposits in southern Missouri. It was widely believed that some of these mystery structures might be associated with igneous activity that could have produced mineable mineral deposits.

Mystery Structures Along the 38th Parallel

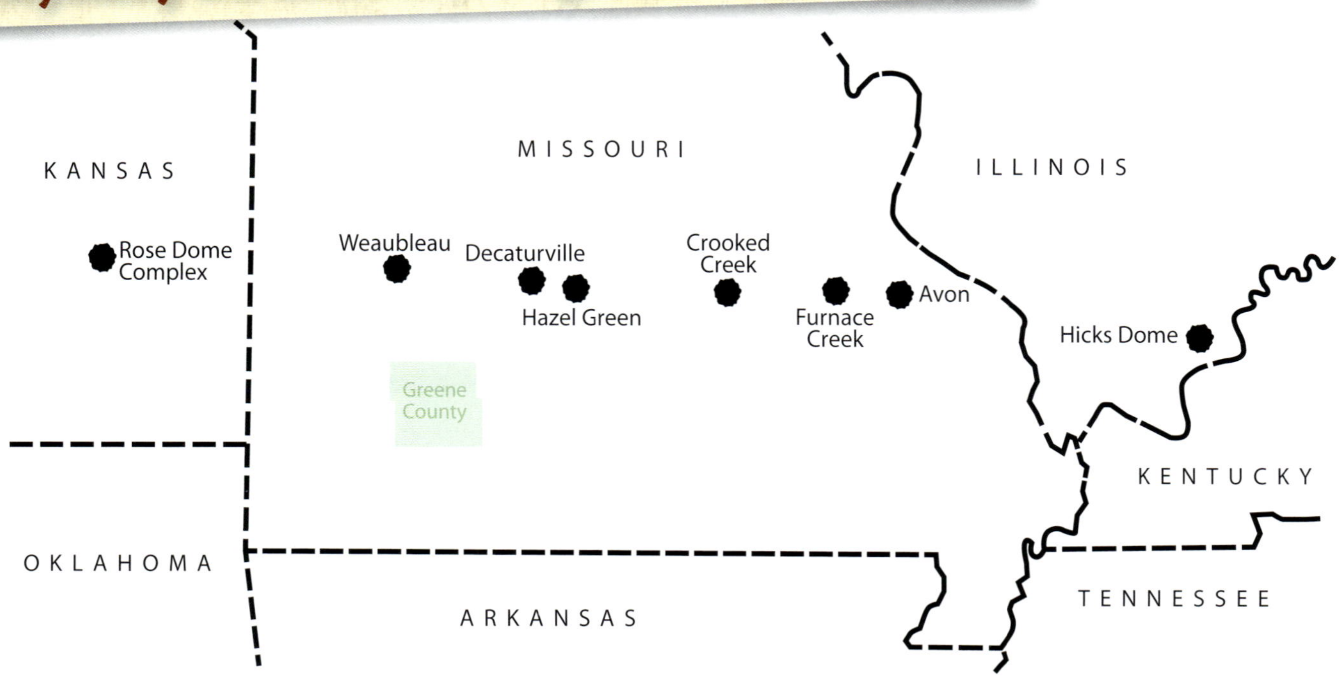

While geologists pondered the origin of the mystery structures, the Space Age began and earth observation satellites began to film the Earth from space and transmit amazing imagery that allowed land-based geologists to see structures that could not be recognized at close range—a geological equivalent of the saying, “can’t see the forest for the trees.”

The space-based imagery gave scientists a whole new perspective on Earth features, but it also brought a new urgency to better understand craters and other features on the Moon. President Kennedy’s drive to place men on the Moon generated intense interest in identifying and studying places on the Earth that might be similar to craters on the Moon. Therefore, the mystery structures along the 38th parallel became prime areas for study.

The Geological Society of America sponsored a field trip during its annual meeting in 1965, featuring the mystery structures in Missouri. The Missouri Geological Survey and the St. Joe Lead Co. provided leaders for the trip, and the Survey published the guidebook, “Cryptoexplosive Structures in Missouri,” a two-day event that visited all of the disturbances except the Rose Dome in Kansas and the Hicks Dome in Kentucky. The field trip attracted participants who thought meteorite impacts were the cause, and others who were convinced the causes were Earth-based.

Spirited debates during the field trip thoroughly explored the evidence available at the time, and the outcome seemed to strengthen the astrobleme camp. Space science was young in the 60s, so there was still strong reliance on traditional

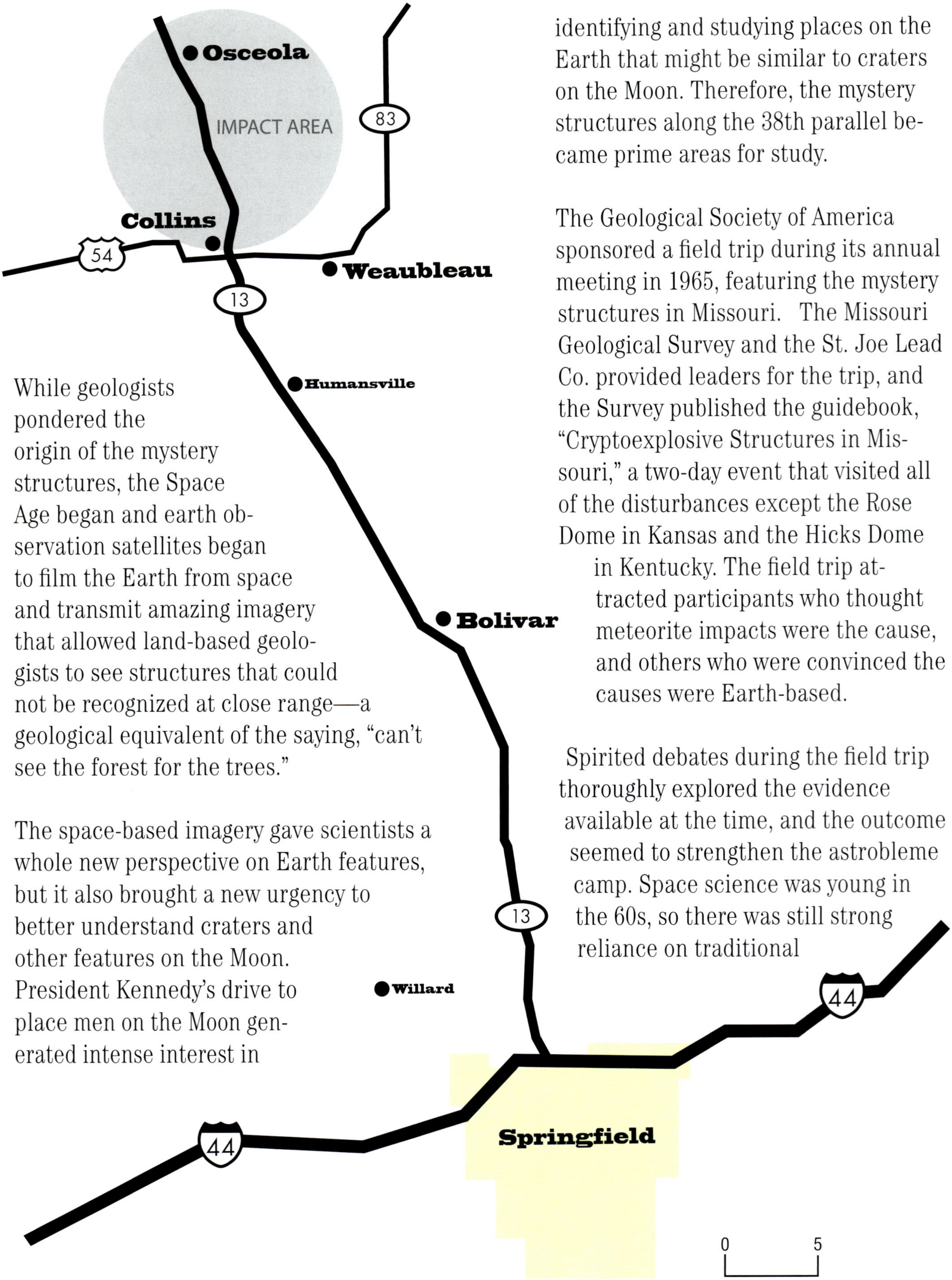

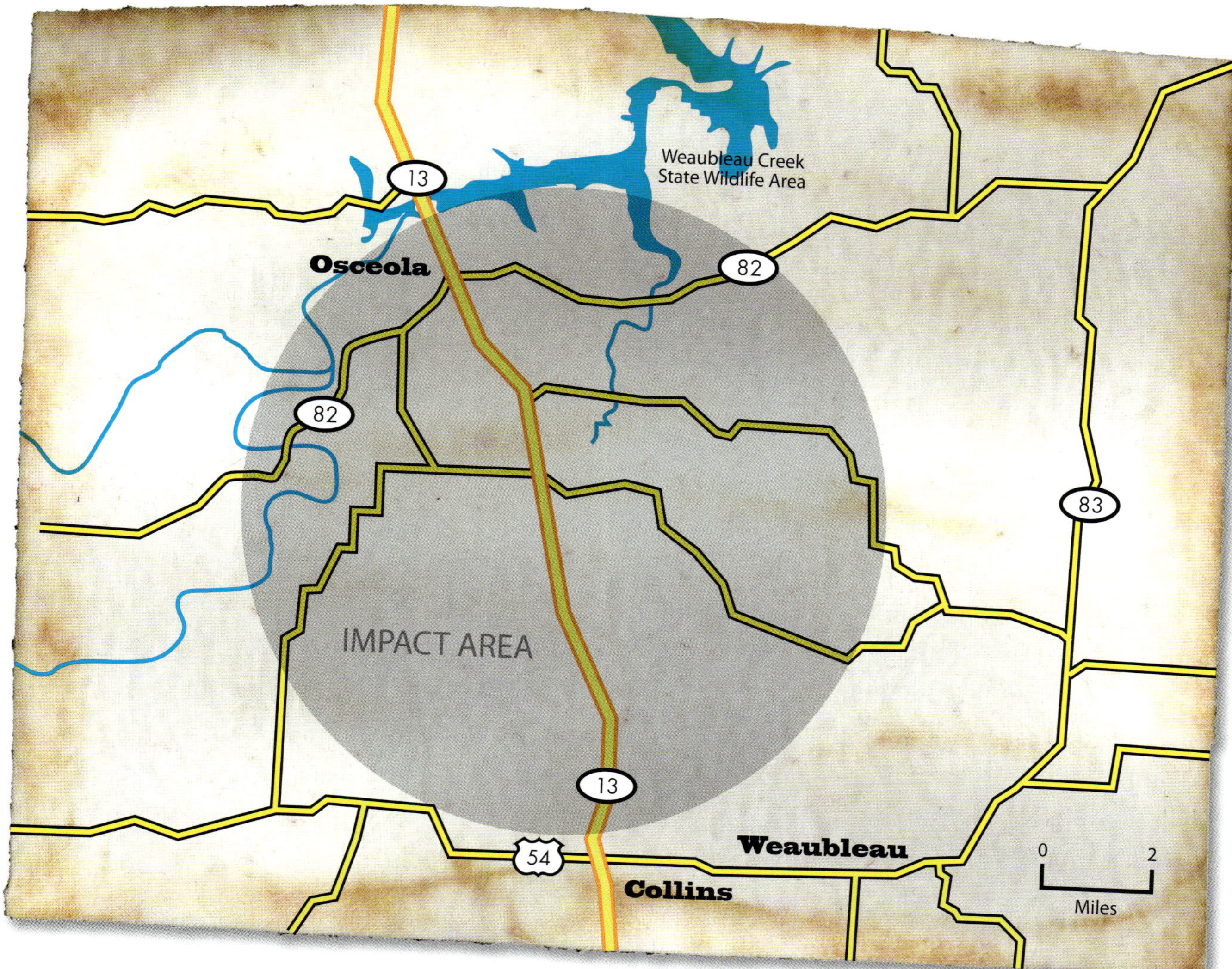

methodology. The guidebook featured an isopach map (isopach maps depict the thickness of geological units) of the Northview formation that showed the unit to be absent in the area later defined as the impact structure for the Weaubleau feature. Moreover, it also showed a poorly-defined area west of the disturbance labeled "Mississippian conglomerate" that later would be considered the debris field for a space object that hit the Earth, creating the Weaubleau astrobleme.

Science often advances through seemingly unrelated events, the Weaubleau structure being a striking example. A quarry was opened by the Ash Grove company in a location that proved to be near the center of the structure. As the quarry grew, it exposed more of the shattered and deformed rocks, and gave geologists a new perspective from which to study the astrobleme.

Geologists have their own version of "Murphy's Law," called the "Edge Effect." For many years geologists have used topographic quadrangle maps as base maps for mapping the geology of an area. In fact, quadrangle mapping is a popular assignment for graduate students working on masters degrees. Tom Beveridge—whose keen sense of humor was legendary—was mapping a quadrangle when he first discovered the Weaubleau Structure. Mappers had a tongue-in-cheek saying, "Anything important will happen on the edge of a quadrangle, and anything extremely important will happen at the corner where four quadrangles intersect." Thus, the edge effect.

When one studies the topography quadrangle by quadrangle, no circular structure is apparent, but when you join the Weaubleau, Vista, Iconium and Osceola 7 ½ minute quadrangles, it is easy to spot a circular structure that is in fact the impact crater caused by an object from space crashing into the Earth. This is essentially what happened when Dr. Kevin Evans, a young faculty member at Missouri State University in Springfield, used a computer to merge what are called "Digital Elevation Models" of the four quadrangles into a single map. Immediately, he recognized a circular structure at the common corners of the four quadrangles. It was a Eureka Moment, and in Evans' words, "Field studies began the next day."

Geologists love field trips, and over time field trips have become popular as ways to share new discoveries among colleagues, and to bring to bear a wider range of skills toward resolving difficult geological questions. So Evans and his colleagues invited the Association of Missouri Geologists to feature the Weaubleau-Osceola area on their 50th anniversary field trip in the fall of 2004.

At the same time, the Mo. Dept. of Transportation was involved in the project to upgrade Hwy. 13 to four lanes, and it happens that the highway crosses the structure slightly west of center. MoDOT had core-drilling equipment on the scene, so they drilled three holes on Hwy. 13 right of way, to give a third dimension perspective to the investigation.

The field trip guidebook laid out the case for defining Beveridge's mystery structure as having been caused by the impact of an object (or objects) from space. So far, no actual impactor has been found, either as a large mass or in fragments, but whatever it was, it left quite a mark. The roughly circular feature is about 10 miles in diameter, with debris (breccia) strewn over a much wider area. The collision created a crater—now worn down by eons of erosion—which penetrated massive layers of limestone and dolomite, with such power that older rock strata were thrust into younger layers, creating the "interthrusts" that so puzzled Tom Beveridge.

The impact must have caused an incredibly powerful explosion, capable of destroying everything nearby and for tens of miles away. The event happened far in the geologic past; had Springfield been in existence at the time, it is likely that buildings would have been flattened and life would have been wiped out.

The story of the Weaubleau-Osceola impact structure is a classic example of how geological science works. Initial discoveries often take years of field and lab work to finally determine how things happened and why. More often than not, the final analysis requires a little serendipity, a little luck, lots of work, and a little help from friends.

Work continues on the Weaubleau-Osceola structure, to determine its age, what kind of object it was, whether it was a single event or part of a serial impact series (such as the comet Shoemaker-Levy 9 that hit the planet Jupiter in 1994), and assessing the risk of similar objects hitting the Earth in modern times.

CHAPTER 11

The Geology–and Politics–of Water Resources

Geology rules; Springfield sits on high ground because of the Burlington Limestone, riddled with sinkholes and graced by springs, a ready source of groundwater for shallow wells.

McDaniel Lake is a vital part of Springfield's water-supply system, which blends surface water, ground-water, springs, and water pumped from Stockton Lake.

Loring Bullard, in his book, *Consider the Source* (2008) documents the fascinating and often contentious story of providing a safe and reliable water supply for the growing Springfield metro area. In the early days, springs provided sufficient water, but as the population grew, water wells had to be drilled and intakes on creeks and rivers had to be developed to meet rising water demands. In time, even these sources were not enough, and Fellows and McDaniel Lakes were constructed, followed by a pipeline to Stockton Lake. With the exception of the pipeline to Stockton Lake, these strategies relied on using water directly available from sources within Greene County.

Toward the end of the 20th century, it became apparent that the region's groundwater and surface water supplies might not be sufficient to meet growing water demands, so planners looked to sources beyond the Greene County boundaries. The most obvious opportunities seemed to lie with the large Corps of Engineers lakes to the south (Table Rock Lake) and northwest (Stockton Lake). Both of these lakes had water-supply storage, thanks to foresightful planning by a little-known state agency called the Water Resources Board, dating back to what has been called the "Big Dam Era." Following the end of World War II, the nation began building large dams as one means of providing employment for veterans as the country changed to a post-war, peacetime economy. Dams were seen as highly beneficial, controlling floods, generating hydropower, enabling river navigation, agricultural irrigation, benefits to fish and wildlife, recreation, and water supply. As a result, many dams were built throughout the nation, in an era of construction that may never be equaled.

Because of the "flood of a century" on the Mississippi River in 1927, the U.S. Army Corps of Engineers became the primary federal agency involved in building dams and levees as well as other water-control structures, mostly in the eastern part of the U.S. (Barry, 1997). The U.S. Bureau of Reclamation also built many dams in the western part of the nation, but its authority was less sweeping than that of the Corps of Engineers. The Missouri Water Resources Board, anticipating future needs, asked the Corps to include water-supply storage as a formal Project Purpose for several of the large dams built in Missouri. This was a brilliant strategy that allowed the State of Missouri to access the enormous storage capacity of federal reservoirs without having to provide the money up-front to construct the dams. The cost of including water-supply storage would have to be repaid, but not until the storage was actually used.

For example, assume that a federal reservoir has Water-Supply Storage as a project purpose, and Springfield wishes to use part of that storage. To understand how it works, imagine a bathtub-size tank, open at the top, with a drain in the bottom and a faucet for filling the tank. The bathtub represents a federal reservoir, such as Table Rock Lake. The tub drain is the water released through the dam, as the White River. The faucet represents the inflow to the lake that keeps it filled. Now imagine going to the hardware store to buy a dipper big enough to meet your water-supply needs. This represents the water-supply storage you have acquired. Notice that you are not buying just a dipper-full of water, but the dipper itself. You may fill and empty the dipper as many times as you like, so long as the inflow to the lake keeps the water levels high enough. Typically, water-supply storage is reliable 98 percent of the time.

This is essentially the procedure followed by the City of Springfield when they completed a pipeline from Stockton Lake to a discharge point in Fellows Lake, where the water is stored and withdrawn as needed to meet the City's water needs.

Springfield and Greene County occupy high ground, astride a major river basin divide that separates waters draining to the Missouri River on the north, from the waters that flow south into the White River basin. This is a comparatively rare situation; most large cities are either on major rivers i.e., Kansas City, or at the confluence of major rivers (St. Louis: Mississippi and Missouri Rivers). Geology rules; Springfield sits on high ground because of the Burlington Limestone, riddled with sinkholes and graced by springs, a ready source of groundwater for shallow wells. The Burlington Limestone is near the top of the geologic column in Greene County, and the county's "footprint" overlies a thick sequence (~2,000 ft.) of predominantly carbonate rocks that function as aquifers. The first aquifer—from the surface down through the rock strata to the Northview

Drainage Map of Greene County

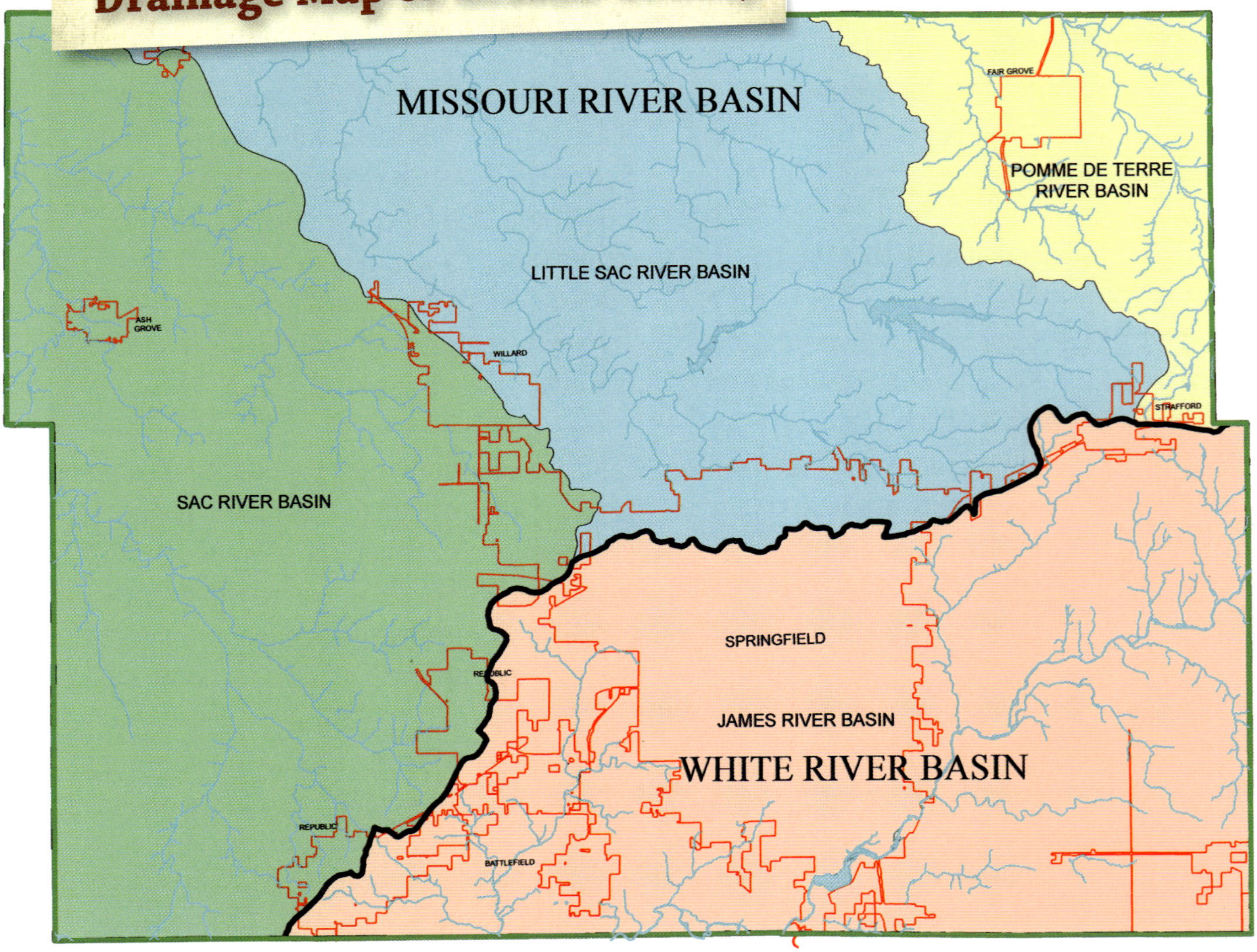

Formation—is called the Springfield Plateau aquifer. Shallow wells drilled into this aquifer yield enough water to meet the needs of individual homes and/or farms and small businesses. The next aquifer is known as the Ozark aquifer; it includes the Ordovician and Cambrian formations down to the Derby-Doe Run and Davis Shale units. Below the Ozark aquifer to the Precambrian basement rocks lies the St. Francois aquifer, consisting of the Bonneterre Dolomite and Lamotte Sandstone. Deep, high-production wells usually penetrate the Ozark aquifer, and—simply put—the deeper the well, the higher the yield.

Springfield's growing demand for water has not gone unnoticed on the national scene. In the 1990s, when the City of Springfield began to import water from Stockton Lake, upstream states in the Missouri River basin objected, arguing that it constituted an inter-basin water transfer, something Missouri had long opposed.

Here is their rationale:

The Missouri River arbitrarily begins at Three Forks, Montana, at the confluence of the Gallatin, the Madison, and the Jefferson Rivers.

However, following the longest of these "parent" streams to its ultimate source leads to a place called Hellroaring Spring, in a remote part of Yellowstone National Park. From this little-known beginning the Missouri River drains parts of 10 states as it winds its way 2,000+ miles to its confluence with the Mississippi River just north of St. Louis. Along the way, its waters are used and/or diverted for many purposes, including domestic and industrial water supply, agricultural irrigation, hydropower generation, navigation, Indian tribes, recreation, flood control, and fish and wildlife. The river is controlled by a series of large dams built and operated by the Corps of Engineers through a cooperative process involving the Corps and the user community that benefits from the river. In recent years, the development of an Annual Operating Plan for the Missouri River has become more contentious, as the various user groups vigorously compete for allocation of an increasingly more limited resource. All this controversy is exacerbated by basic geologic and geographic differences in various parts of the Missouri River basin: States in the upper basin—upstream states—have water laws based on the principle of "first in time, first in right," or what has become known as the doctrine of prior appropriation. Downstream states generally have less restrictive water laws, or none at all. Missouri uses the riparian doctrine, which follows the principle of "reasonable use;" in effect, Missouri has NO water rights legislation.

Because of its favorable geographic location, Missouri benefits from adequate annual precipitation and thus far has been able to manage its water resources without the constraints of Western water law. Of course there are occasional issues that must be settled in the courts, but up to this point is has not been necessary to enact strict water laws, establish "water courts," and fund the necessary bureaucracy to carry out policies that will safeguard our access to water for future generations. As Mark Twain so aptly put it, "Whiskey is for drinkin,' water is for fightin.'"

Out-of-basin water diversion is easily the most divisive issue in the Missouri River basin, one that has already been raised in regard to Springfield's tapping Stockton Lake for additional water supply. The Garrison Diversion Project in North Dakota typifies the diversion issue. The Missouri River begins in the mountains of northwestern Wyoming, flows through the vast wheatfields of northern Montana, and into North Dakota just as it is swelled by the Yellowstone River as it enters the Missouri west of Williston. From there it flows across the southwestern quarter of the state, until it crosses the border into South Dakota, south of Bismarck, North Dakota's capitol. The Corps of Engineers built massive Garrison Dam some 60 miles NNW of Bismarck, impounding Lake Sacajawea, one of the largest reservoirs in the Missouri River system. This lake is so vast that it resembles an inland sea in a region of rolling prairies that stretch to the horizon, the fabled wide-open spaces of the American West.

Decades ago, the state of North Dakota devised a plan to divert water from the Missouri River to provide a reliable water supply for the water-short northeastern part of the state. To gain needed political support, they included parts of northwestern Minnesota as well, and threw in the flooding problems at Devils Lake, North Dakota, as a bonus. This plan would take water from Lake Sacajawea and pump it through a series of canals and small reservoirs along the Sheyenne River valley, to Devils Lake and eventually into the Red River of the North, which drains into Canada and the Arctic Ocean.

The plan became known as the Garrison Diversion Project, and it met immediate and intense resistance from downstream states, who saw it as diminishing the flow of the Missouri and Mississippi Rivers. Environmentalists also attacked the plan, seeing great risk in opening a route for invasive species to move from the Missouri River, through the Garrison Diversion, into Hudson Bay. Through years and years of controversy, the Garrison Diversion Project stayed alive, as the system was built element by element, using funds appropriated by Congress. North Dakota is a low-population, mostly rural state, unable to afford massive water projects, so its representatives in Congress were extremely effective in winning Congressional approval for a series of incremental parts of the project. The largest element was a massive pumping station near Garrison Dam, designed to take water from the Missouri River and pump it upstream, eventually out of the Missouri River basin. Today the project is on the verge of completion, another landmark in the long and contentious history of the Missouri River basin.

There have been numerous other schemes for diverting water from the Missouri River. Although these may have laudable goals, the end result is less water flowing down the Missouri into the Mississippi River, and ultimately into the Gulf of Mexico. One of the most persistent plans involves diverting water from the Missouri River and using it to recharge the Ogallala groundwater aquifer to support massive irrigation projects in Nebraska, Kansas, Oklahoma and Texas. While large-scale diversions to recharge the aquifer have not yet been completed, the pumping of groundwater continues and may intensify in the years ahead, as the aquifer becomes further depleted.

Perhaps the most unusual scheme for using water diverted out-of-basin from the Missouri River would have used the river water to operate a coal-slurry pipeline to move low-sulfur coal from surface mines in Montana and Wyoming, to power plants in the Southwest. Coal resources in the central part of the U.S. usually have high sulfur content; when the coal is burned as fuel for an electrical generating plant, the sulfur released as stack gases becomes the source for acid rain, one of the most detrimental effects of using coal for power generation. However, coal in Montana and Wyoming is low-sulfur, and it is more cost-effective to move coal by train from those states and use sulfur-removal technology at the power plants, than it is to use high-sulfur Missouri coal.

In order to cut the cost of shipping coal by train for hundreds of miles, engineers conceived the idea of pulverizing the coal at the mine, mixing it with water from the Missouri River to make a slurry, and pumping it to power plants far to the south. A pilot-size facility was actually built near Columbia, Missouri, to test the feasibility of the idea, which proved workable. However, the financing for a full-scale pipeline failed to materialize, and yet another diversion idea faded into obscurity.

Keeping up with Springfield's growing demands for water has always been a challenge, one that is likely to intensify in the future. One need only look at post-Katrina Louisiana to see why. Several decades ago, one of the staples of geography was the idea that the delta of the Mississippi River—often said to resemble a bird's foot and therefore called the "Birdfoot Delta"—was continually expanding southward into the Gulf of Mexico. Eventually, as students were taught, it would gradually fill in the Gulf, and the Port of New Orleans would become an inland port. At some point, though, the growth ceased and the delta began to recede. Currently, the State of Louisiana is losing about 29 square miles of its coastal lands each year, as the Birdfoot Delta endures a "perfect storm" of negative resource decisions that will be difficult to reverse.

Consider:

1. The construction of dams on the upper Missouri River and the Mississippi River above St. Louis has increased the loss of water by evaporation, equivalent to a diversion.
2. The dams interrupt the downstream movement of sediments, which instead of providing the materials to increase the growth of the delta, are stored behind the dams.
3. Increased consumptive use of water throughout the major river basins means gradual reduction in the cumulative flows of the rivers.

The cave stream flowing through Fantastic Caverns surfaces at the base of a bluff along the Sac River. The flow in dry seasons is a trickle compared to the powerful surge following heavy, sustained rainfall.

It seems inevitable that, as water resources decline in the face of increased demand, there will be more contentious debates over who gets to use how much water from what sources. Missouri, which operates under the riparian (reasonable use) doctrine of water law, will find it increasingly difficult to justify allowing its water users to withdraw as much water as they wish. Water conservation is an imperative that cannot be ignored.

Groundwater

Under Greene County's "footprint" there is a remarkable groundwater resource, contained in the sedimentary rock formations between the land surface and the Precambrian igneous and metamorphic rocks of the "basement." The proof of this is in the hundreds of springs that dot the landscape with clear, cold, life-giving water. This groundwater is a renewable resource that is replenished—or recharged—by that part of the annual precipitation that is not lost to evaporation or drained away by surface streams. The Natural Well—commemorated by a monument on Water Street across from Founders Park in downtown Springfield–was a natural opening into what is now known as the Springfield Plateau aquifer, or simply "the shallow aquifer." [An aquifer is a rock formation capable of storing and yielding groundwater.]

Imagine a block of rock in the shape of Greene County, 2,000 feet thick. That block represents the part of the Earth's crust that holds—in the natural pore space in the rock formations—all of the groundwater available for use in Greene County. The fact that springs drain the upper part of this block seems counterintuitive, because it is commonly believed that "the water table is dropping," so how is it possible for springs to continue flowing? The answer lies in the region's karst topography, developed on surface rocks that are primarily limestone. Over time, underground drainage networks develop by solution of the bedrock, and these networks—many of them cave systems that can be explored and mapped—sustain the flow of springs even as groundwater withdrawal through water wells causes cones of depression to form in the local water table. In karst areas, the concept of a water "table" becomes much more complicated than it is in areas where landscapes develop over non-soluble rocks. The sedimentary rock units under Greene County's "footprint" are saturated with groundwater from the water table down to the Precambrian "basement." The igneous and metamorphic rocks of the basement are not considered to be aquifers, so the groundwater available for use is all contained in the 2,000 feet thick block of sedimentary rocks resting on the basement. These rock formations have been grouped into three productive aquifers:

1. **The Springfield Plateau aquifer**, consisting of the Warsaw, Burlington-Keokuk, Elsey, and Pierson formations;
2. **The Ozark aquifer**, consisting of the Cotter, Jefferson City, Roubidoux, Gasconade, Eminence, and Potosi formations; and
3. **The St. Francois aquifer**, consisting of the Bonneterre Dolomite and the Lamotte Sandstone. These major aquifers are separated by confining units—rock formations that prevent groundwater from moving through them. The confining unit between the Springfield Plateau aquifer and the Ozark aquifer consists of the Northview, Compton, and Bachelor formations. The Derby-Doe Run and Davis formations separate the Ozark and St. Francois aquifers.

The Springfield Plateau aquifer is the water source for thousands of shallow wells in Greene County, ranging from hand-dug wells to drilled wells of various depths. The maximum thickness of the aquifer is about 400 feet. Wells typically yield 10 to 30 gallons per minute. In a perfect world, every well would be drilled properly, casing would protect from contamination, and well water would be as good today as it was when the first settlers began to populate the area. Unfortunately, we live in a world that is far from perfect, so well users must be alert for contaminants in their water supply.

The Ozark aquifer yields high-quality groundwater in copious quantities. It consists mostly of thick dolomite and some sandstone rocks with an aggregate thickness of about 1,000 feet. Wells drilled through the full thickness of the aquifer may yield well over 1,000 gallons per minute. High-production wells drawing from the Ozark aquifer typically have casing that seals off contaminant sources.

The Department of Natural Resources (DNR) maintains a well information management system that contains data on wells drilled from 1987 to date. Up to 1987, there was no legislation regulating well drilling, and no requirements for casing to prevent aquifer contamination. However, following passage of enabling legislation well drillers were required to demonstrate competency and to meet standards for well construction and casing. They were required to certify that wells were drilled in compliance with accepted standards, and each well drilled was reported to DNR.

Requirements are also in place for the capping of abandoned wells and mineral test holes, to insure that open drillholes do not provide pathways for contaminants to move from the surface into underlying aquifers.

Determining precisely how many water wells exist in Greene County is difficult, because prior to 1987 it was not necessary to report drilling statistics to DNR, although many wells were documented. There are records on 1,031 wells that were drilled prior to 1987, plus an additional 5,216 wells documented since 1987. This is a grand total of 6,247 well records in Greene County.

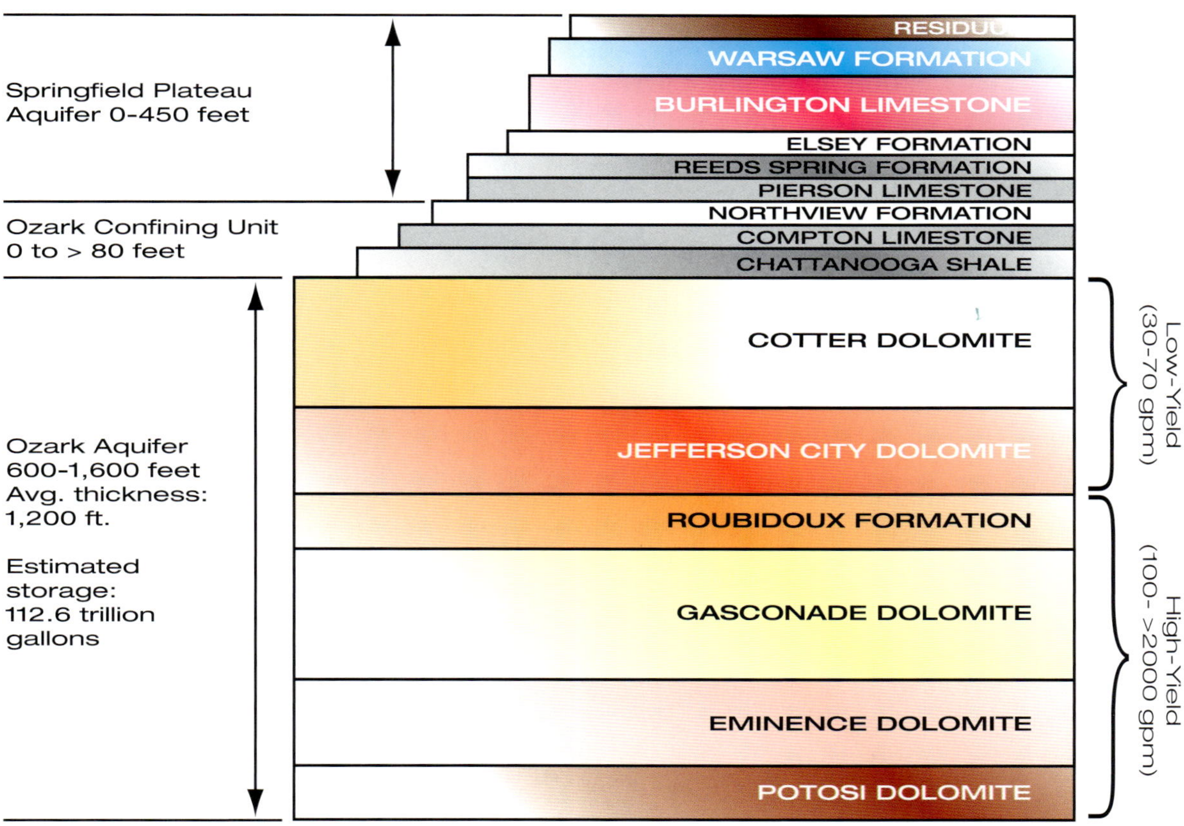

Acknowledgements

"The Hood Ornament" gargoyle, Hwy. 65 south.

Writing this book was a learning adventure, made more pleasurable by the generous ways in which so many people helped me with information, access to properties, photos, maps, and encouragement. Loring Bullard suggested the project and his determination to see it completed kept the work in progress through ongoing budgetary uncertainty. Loring and his staff at the Watershed office were unfailingly helpful, creative, and energetic. Stacey Armstrong, skilled in GIS work, did several of the maps. Eric Baker of Blue Sky Design did layout and design work for the book, and produced several line drawings as well.

Louis Griesemer and David Holberg of Springfield Underground, Chris Upp and Dave Sturdevant of Conco Quarries, and Dave Richter and Freddie Flores of Phenix Quarry provided access and photos of limestone producing operations in Greene County. Jack Ray of M.S.U.'s Center for Archaeological Research shared his knowledge of metal mining sites.

Employees of both the City of Springfield and Greene County generously shared their time and expertise as they provided access to publicly-owned sites. I thank Roddy Rogers of City Utilities, Melvin Johnson of Parks, and Mark Ryan of Public Works, who were especially helpful. Jim Vandiver of Greene County GIS produced several maps using the facilities of the county's Geographic Information System.

Jon Beard, Roy Gold, and Charley Young of the Springfield Plateau Grotto of the National Speleological Society provided cave information and offered photos to help document the rich cave resources of the area. Matt Forir of the Missouri Institute of Natural Science offered information and photos of Riverbluff Cave.

Wherever I traveled throughout the city and county, I found pleasant, knowledgeable people who were eager to share what they knew of the place we call our "home country." The area's fascinating geology is clearly a primary reason why this is such an inspiring place to live and work. So many people assisted me that I may have failed to mention some, though any omissions were not intentional.

Writing the book took far longer than I had promised, but through many hours of computer work, field work, photography and research, my wife Helen was ever supportive and encouraging, though piles of maps, rock samples, and scattered manuscript pieces severely tried her patience.

Funding support for this book provided by the Southeast Rotary Club of Springfield, Greene County, City of Springfield Department of Public Works Stormwater Division, Springfield Underground, and Watershed Committee of the Ozarks.

About the Author

There's an old Rock n' Roll song with the line, "don't know much about history." Perhaps the next line should've been, "don't know much GE-OLO-GY," because it, like physics or trigonometry, is a foreign subject to most people, who typically have only vague inklings about the underlying geology where they live, work, build and play. Once in a while a guy like Jerry Vineyard comes along, who can take a complicated, technical topic like geology and make it accessible and understandable to the "average" citizen. That is a rare and important talent, the satisfying results of which are found in this book.

Author, left, with Ken Thomson.

Jerry grew up in Dixon, Missouri, in the northern part of the Ozarks, where at an early age he took a keen interest in the rocks and arrowheads and springs on his family's farm. He explored many of the caves along the nearby Gasconade River, motivating him to create a list of the area's known caves. His research impressed Tom Beveridge, Missouri's State Geologist at the time, who gave Jerry his first real job, even providing him with a Jeep to more widely explore and catalog cave resources. Jerry still laughs when he recalls, "I didn't think someone could get an actual job doing this stuff."

That actual job, in fact, led to further employment in the field and many important accomplishments, as Jerry went on to become Program Director for the Missouri Geological Survey as well as a writer, teacher and river negotiator. He tackled huge and complex management issues surrounding the Missouri-Mississippi River Basins. As principal negotiator for the state in discussions with the Army Corps of Engineers and the Clarence Cannon Wholesale Water Commission, he obtained a century-long guarantee for a safe, reliable water supply from Mark Twain Lake in northeast Missouri—one of his proudest moments.

Jerry performed his first major geological work in Greene County in the mid-1960s, when he became involved with water pollution issues in Wilson's Creek, the stream receiving most of the treated sewage from the city of Springfield. This was before EPA, so Jerry worked with the federal Water Pollution Control Administration to ensure that pollution problems in Wilson's Creek were addressed before the creation of Wilson's Creek National Battlefield Park, located on the stream just below the wastewater plant.

With his life-long passion for subterranean geology, Jerry spent a good portion of his career underground, crawling and exploring and mapping his way through hundreds of caves all over the state. He proudly recalls when, in 1990, Missouri became the first state in the nation to record 5,000 caves and Governor Ashcroft declared the "Year of the Cave." With co-author Gerald Feder, Jerry produced *Springs of Missouri*, a beautiful book gracing the shelves of home and business libraries around the world. Missouri Life magazine referred to Jerry as "Missouri's pre-eminent geologist/explorer." But to many of us, he is simply a respected and trusted scientific veteran, our "caveman" and "spring guy."

Jerry is retired now, and lives with his wife Helen and their cat (Muffin) in Ozark, Missouri. Retirement, of course, is a relative state, and Jerry still logs many hours as a volunteer, serving for example on the Board of Directors of the L-A-D Foundation, owner of Pioneer Forest, a 140,000 acre sustainable working forest near the Current River. He chairs the organization's Grants Committee, which uses some of the forest earnings to fund projects benefiting the natural and cultural resources of the Ozarks.

Even with his post-retirement activities, Jerry was enjoying a rather peaceful existence in Christian County until the Watershed Committee of the Ozarks discovered his whereabouts. What better person, we asked ourselves, to update the 1980s edition of the "Geology of Greene County," originally authored by SMSU (now MSU) geology professor Ken Thomson? Jerry, however, could never be content with just "updating the information." Instead, he has produced a work that would've made Ken Thomson very proud, because it is an informative but easy read—it brings geology to the people. We know that you will enjoy your copy of "Gargoyles," and hope that it will cause you to explore, and appreciate, the inspiring geology of Greene County.

Loring Bullard,
Executive Director
Watershed Committee of the Ozarks

In Memory of Ken Thomson

In the Ozarks, particularly Greene County, it is an interesting fact that many, if not most folks, know what you're talking about when you use the word "karst". One would think that a morphologic concept as complex as karst would be unfamiliar at least, or misunderstood at best. However, thanks to the educational commitment of many dedicated scientists, and one in particular, the local population, for the most part, "gets it!"

Recognizing the importance of karst to southwest Missouri, and the need to elevate our collective knowledge of karst systems, a geology professor from Missouri State University (MSU) stepped forward in the late 1960s. He made it his mission to educate the citizens of southwest Missouri about the geologic processes that give birth to caves, springs, and sinkholes.

Kenneth C. Thomson, PhD., served as a Professor of Geology at MSU from 1968 through 2004. He was widely known for his trailblazing geologic research and precise mapping skills. He did some of his best work alongside hydrogeologist Tom Aley (owner of Ozark Underground Laboratory), with whom he shared a long and productive professional relationship. Together they published the results of several groundbreaking geologic investigations including:

- Solution Features of Greene County, Missouri (1978)
- Identification of Areas in Unincorporated Greene County where Sinkhole Flooding and Serious Groundwater Contamination Could Result from Land Development (1980)
- Septic Fields and the Protection of Groundwater Quality in Greene County, Missouri (1984)
- Recharge Area Delineations for Twelve Spring Systems, Greene County, Missouri (2002)

Ken and Tom complemented each other. While Tom was the "front man," always ready with a quote or a quip for the news media, "Doc" Thomson preferred to work behind the scenes. He was an excellent illustrator and cartographer. Anyone who has ever utilized a rapidograph inking set to create a map or draw a geologic column knows the skill required and attention to detail necessary to produce an accurate representation. Doc's graphics were impeccable. He worked quickly and largely error free.

Ken passed from life far too quickly in 2004. He will be remembered as a man who spent much of his time doing things for other people. When an F3 tornado hit southwest Springfield in 1983 and damaged his home, he was so busy helping neighbors make repairs that all the roofing supplies were sold out by the time he got to the hardware store. Fortunately, Tom Aley had surplus roofing material and came to Doc's rescue.

As a professor, he made himself available at any time, day or night, to help his students. He had incredible patience and perseverance. His leadership allowed him to overcome obstacles and "rally the troops" to accomplish the task at hand. Whether dealing with a raccoon infested cave or leading the exploration of the oldest known fossil cave in North America (Riverbluff Cave in Greene County), Doc Thomson always accepted the challenge with good humor and good science.

David L. Coonrod
Former Presiding Commissioner
Greene County, Missouri

Sources, And For Further Reading

Barry, John M., 1997, *Rising Tide, The Great Mississippi Flood of 1927 and How it Changed America*: Simon and Schuster, New York

Bretz, J Harlen, 1956, *Caves of Missouri:* Missouri Geological Survey and Water Resources, Vol. XXXIX, Second Series, Rolla, Missouri.

Bullard, Loring, 2005, *Consider the Source: A History of the Springfield,* Missouri Public Water Supply: Watershed Press, Springfield, Missouri.

Bullard, Loring, Kenneth C. Thomson, and James E. Vandike, 2001, *The Springs of Greene County:* Missouri Dept. of Natural Resources, Geological Survey and Resource Assessment Division, Water Resources Rept. No. 68, Rolla, Missouri.

Cooley, Robert E., and Michael Fuller, 1976, *An Archaeological and Historical Survey of Areas to be Affected by the Pierson Creek Trunk Sewer*, *City of Springfield, Greene County, Missouri*: Center for Archaeological Research, Southwest Missouri State University, Springfield, Missouri, Project: CAR-10.

Emmett, Leo F., et al., 1978, *Water Resources and Geology of the Springfield Area, Missouri*: Missouri Dept. of Natural Resources, Division of Geology and Land Survey, Water Resources Report No. 34, Rolla, Missouri.

Evans, Kevin R, Kevin L. Mickes, Charles W. Rovey II, and George H. Davis, 2003, Field Trip I: *The Weaubleau-Osceola Structure: Evidence of a Mississippian Meteorite Impact in Southwestern Missouri*: in Association of Missouri Geologists, 2003 Field Trip Guidebook, Springfield, Missouri.

Fellows, L.D., 1970, *Geology of the Galloway Quadrangle, Greene County, Missouri*: Missouri Geological Survey and Water Resources, Geologic Quadrangle Map Series No. 5, (one sheet plus separate text), Rolla, Missouri.

Ferrell, John R., 1993, *Big Dam Era: A Legislative and Institutional History of the Pick-Sloan Missouri Basin Program*: Missouri River Division, U.S. Army Corps of Engineers, Omaha, Nebraska.

Hawker, Jon L., 1992, *Missouri Landscapes: A Tour Through Time*: Missouri Dept. of Natural Resources, Division of Geology and Land Survey, Educational Series No. 7, Rolla, Missouri.

Tan dolomite landscape stones, dark metal artwork, and recirculated water comprise a welcome display at The Lakes at Wild Horse subdivision, east Division Street. Geology is the centerpiece of many landscape structures throughout Springfield and Greene County.

Hayes, William C., and Jerry D. Vineyard, 1969, *Environmental Geology in Towne and Country*: Missouri Geological Survey and Water Resources, Educational Series No. 2, Rolla, Missouri.

Kisvarsanyi, Eva B., 1974, *Operation Basement: Buried Precambrian Rocks of Missouri—Their Petrography and Structure*: American Association of Petroleum Geologists Bulletin, Vol. 58, No. 4, p. 674-684.

Kisvarsanyi, Eva B., 1975, *Data on Precambrian in Drillholes of Missouri, Including Rock Type and Surface Configuration*: Missouri Dept. of Natural Resources, Geological Survey, Report of Investigations No. 56, Rolla, Missouri.

Miller, Don E., and James E. Vandike, 1997, *Groundwater Resources of Missouri*: Missouri Department of Natural Resources, Division of Geology and Land Survey, Volume II, State Water Plan Series, Rolla, Missouri.

Park, John R., 2005, *Missouri Mining Heritage Guide*: Stonerose Publishing Co., South Miami, Florida.

Snyder, Frank G., James H. Williams, et al., 1965, *Cryptoexplosive Structures in Missouri*: Missouri Geological Survey and Water Resources, Report of Investigations No. 30, Rolla, Missouri.

Thomson, Kenneth C., 1986, *Geology of Greene County, Missouri*: Watershed Management Coordinating Committee, Springfield, Missouri.

Thompson, Thomas L., 1986, *Paleozoic Succession in Missouri: Part 4, Mississippian System*: Missouri Department of Natural Resources, Division of Geology and Land Survey, Report of Investigations No. 70, Rolla, Missouri.

Thompson, Thomas L., 1995, The *Stratigraphic Succession in Missouri*: Missouri Department of Natural Resources, Volume 40 (Second Series), Revised, Rolla, Missouri.

Thompson, Thomas L., and Charles E. Robertson, 1993, *Guidebook to the Geology Along Interstate 44 (I-44) in Missouri*: Missouri Dept. of Natural Resources, Division of Geology and Land Survey, Report of Investigations No. 71, Rolla, Missouri.

Unklesbay, A.G., and Jerry D. Vineyard, 1992, *Missouri Geology: Three Billion Years of Volcanoes, Seas, Sediments, and Erosion*: Univ. of Missouri Press, Columbia, Missouri.

Vandike, James E., and L. Daniel Sheman, 1994, *Hydrogeologic Investigation of the Fulbright Area, Greene County, Missouri*: Missouri Dept. of Natural Resources, Division of Geology and Land Survey, Water Resources Report No. 43, Rolla, Missouri.

Vineyard, Jerry D., and Gerald L. Feder, 1982, *Springs of Missouri: Missouri Dept. of Natural Resources*, Division of Geology and Land Survey, Water Resources Report No. 29, Rolla, Missouri.

Vineyard, Jerry D., 1997, *Water Resource Sharing: The Realities of Interstate Rivers*: Missouri Dept. of Natural Resources, Division of Geology and Natural Resources, Missouri State Water Plan Series, Vol. VI, Water Resources Report No. 50, Rolla, Missouri.

Vineyard, Jerry D., and Larry D. Fellows, 1967, *Guidebook to the Geology Between Springfield and Branson, Missouri, Emphasizing Stratigraphy and Cavern Development*: Missouri Geological Survey and Water Resources, Report of Investigations No. 37, Rolla, Missouri.

Waite, Loyd A., and Kenneth C. Thomson, 1993, *Development, Description, and Application of a Geographic Information System Data Base for Water Resources in Karst Terrane in Greene County, Missouri*: U.S. Geological Survey, Water Resources Investigation Report 93-4145, Rolla, Missouri.

Weaver, H. Dwight, and Paul A. Johnson, 1980, *Missouri the Cave State: Dark Pathways Series, Book 4*, Discovery Enterprises, Jefferson City, Missouri.

Wharton, Heyward M., 1969, *Missouri Minerals: Resources, Production, and Forecasts*: Missouri Geological Survey and Water Resources, Special Publication No. 1, Rolla, Missouri.

Vuggy Jefferson City Dolomite.